B-47 STRATOJET

Walter J. Boyne Military Aircraft Series

F-22 Raptor
America's Next Lethal War Machine
STEVE PACE

B-1 Lancer
The Most Complicated Warplane Ever Developed
DENNIS R. JENKINS

B-24 Liberator
Rugged but Right
FREDERICK A. JOHNSEN

B-2 Spirit
The Most Capable War Machine on the Planet
STEVE PACE

F/A-18 Hornet
A Navy Success Story
DENNIS R. JENKINS

B-17 Flying Fortress
The Symbol of Second World War Air Power
FREDERICK A. JOHNSEN

F-105 Thunderchief
Workhorse of the Vietnam War
DENNIS R. JENKINS

B-47 Stratojet
Boeing's Brilliant Bomber
JAN TEGLER

B-47 STRATOJET

Boeing's Brilliant Bomber

Jan Tegler

McGraw-Hill

New York   San Francisco   Washington, D.C.   Auckland   Bogotá
Caracas   Lisbon   London   Madrid   Mexico City   Milan
Montreal   New Delhi   San Juan   Singapore
Sydney   Tokyo   Toronto

Library of Congress Cataloging-in-Publication Data

Tegler, Jan
    B-47 stratojet : Boeing's brilliant bomber / Jan Tegler.
        p.   cm.
    Includes index.
    ISBN 0-07-135567-7
    1. B-47 bomber.   2. United States. Air Force. Strategic Air
Command.   I. Title.

    UG1242.B6 T43 2000
    623.7'463'0973—dc21

                                            00-037256

McGraw-Hill
A Division of The **McGraw·Hill** Companies

    2 3 4 5 6 7 8 9 0   QK/QK   0 9 8 7 6 5 4 3 2 1

ISBN 0-07-135567-7

The sponsoring editor for this book was Shelley Ingram Carr, the editing supervisor was Caroline Levine, and the production supervisor was Sherri Souffrance. It was set in Utopia by North Market Street Graphics.

Printed and bound by Quebecor/Kingsport.

 This book is printed on recycled, acid-free paper containing a minimum of 50% recycled, de-inked fiber.

McGraw-Hill books are available at special quantity discounts to use as premiums and sales promotions, or for use in corporate training programs. For more information, please write to the Director of Special Sales, McGraw-Hill, 11 West 19 Street, New York, NY 10011. Or contact your local bookstore.

The McGraw-Hill Companies is pleased to present the **Walter J. Boyne Military Aircraft Series.** The series will feature comprehensive coverage, in words and photos, of the most important military aircraft of our time.

Profiles of aircraft critical to defense superiority in World War II, Korea, Vietnam, the Cold War, the Gulf Wars, and future theaters, detail the technology, engineering, design, missions, and people that give these aircraft their edge. Their origins, the competitions between manufacturers, the glitches and failures and type modifications are presented along with performance data, specifications, and inside stories.

To ensure that quality standards set for this series are met volume after volume, McGraw-Hill is immensely pleased to have Walter J. Boyne on board. In addition to his overall supervision of the series, Walter is contributing a Foreword to each volume that provides the scope and dimension of the featured aircraft.

Walter was selected as editor because of his international preeminence in the field of military aviation and particularly in aviation history. His consuming, lifelong interest in aerospace subjects is combined with an amazing memory for facts and a passion for research. His knowledge of the subject is enhanced by his personal acquaintance with many of the great pilots, designers, and business managers of the industry.

As a Command Pilot in the United States Air Force, Colonel Boyne flew more than 5000 hours in a score of different military and civil aircraft. After his retirement from the Air Force in 1974, he joined the Smithsonian Institution's National Air & Space Museum, where he became Acting Director in 1981 and Director in 1986. Among his accomplishments at the Museum were the conversion of Silver Hill from total disarray to the popular and well-maintained Paul Garber Facility, and the founding of the very successful Air&Space/Smithsonian magazine. He was also responsible for the creation of NASM's large, glass-enclosed restaurant facility. After obtaining permission to install IMAX cameras on the Space Shuttle, he supervised the production of two IMAX films. In 1985, he began the formal process that will lead ultimately to the creation of a NASM restoration facility at Dulles Airport in Virginia.

Boyne's professional writing career began in 1962; since that time he has written more than 500 articles and 28 books, primarily on aviation subjects. He is one of the few authors to have had both fiction and nonfiction books on the New York Times best seller lists. His books include four novels, two books on the Gulf War, one book on art, and one on automobiles. His books have been published in Canada, Czechoslovakia, England, Germany, Italy, Japan, and Poland. Several have been made into documentary videos, with Boyne acting as host and narrator.

Boyne has acted as consultant to dozens of museums around the world. His clients also include aerospace firms, publishing houses, and television companies. Widely recognized as an expert on aviation and military subjects, he is frequently interviewed on major broadcast and cable networks and is often asked by publishers to review manuscripts and recommend for or against publication.

Colonel Boyne will bring his expertise to bear on this series of books by selecting authors and titles and working closely with the authors during the writing process. He will review completed manuscripts for content, context, and accuracy. His desire is to present well-written, accurate books that will come to be regarded as definitive in their field.

As author Jan Tegler points out in his excellent text, the Boeing B-47 Stratojet has been largely overlooked by historians despite the aircraft's seminal importance to aviation history. While hundreds of books have been done on World War II fighters, such as the Mustang or the Spitfire, only a very few have been devoted to the B-47. Tegler rectifies the situation with a comprehensive overview of the handsome, six-jet bomber, providing a mammoth amount of data and a wealth of photos.

Some explanations have been made in the past about the lack of coverage given to the B-47, including the fact that it never served in combat and was never given a nickname. There is some validity in this for warplane reputations are ordinarily best made in combat. One can only thank God that the B-47 never had a chance to earn such a reputation, for it was designed for nuclear warfare. The fact that the great numbers of B-47s undoubtedly deterred the Soviet Union is tremendously important—but not something to excite authors. And it is true that nicknames are invaluable in creating an image about an aircraft. As quick examples, the Douglas C-47 was immortalized as the Gooney Bird and the Boeing B-52 as the BUFF.

Flying and maintaining the B-47 was such a serious business that the aircrews and the ground personnel shied away from a nickname, and only public affairs people called it the "Stratojet." When it appeared on the scene it was new, powerful, and laden with the danger associated with innovation and high performance. Throughout its more than a decade of service it remained a "hot" airplane, one to be regarded with respect and affection.

Although there were some dissenters, most of the people who flew the B-47 came to love the airplane, simply because it performed so well and so responsively. For pilots brought up on Boeing B-17s and B-29s, the high performance of the B-47 was intoxicating. It was wonderful to be able to *climb* at 310 knots and break out on top of the weather, after years of nosing up at 120 knots or less, and then bore through the clouds and turbulence. And while the aircraft did demand close control of airspeed in all regimes of flight, it was not difficult to do so—a mere caress of the throttles could add or subtract from the airspeed, a knot at a time.

But there was something else that was perhaps even more important. When the B-47s began entering operational service, the Strategic Air Command was just coming into its own as the premier organization not only in the Air Force, but among all the armed forces of the United States. It was clearly the first team, the deterrent force upon which the defense of the free world was based. It had reached this status because of the will, planning, and determination of General Curtis E. LeMay who set high standards and then demanded that everyone live up to them.

Life in SAC was often difficult, especially for the families, and the divorce rate was high. It was not uncommon to fly 80 or 100 hours a month, and temporary tours of duty away from home were both frequent and unpredictable. No family requirement, not even the birth of a baby or a child's illness, was allowed to interfere with the flying schedule.

Despite these difficulties, it was evident during the early years of the B-47 that it was a superb offensive weapon, one with which the Soviet Union could not cope. Through the late 1950s, it was apparent that a Soviet invasion of Europe could be immediately countered by a massive attack by B-47s that would have destroyed the enemy homeland with relatively few losses. The 1960s would see a change in this, of course, with the advent of both surface-to-air and ballistic missiles.

The combination of the aircraft's high performance and its vital mission made it immensely satisfying to fly. In its early years, the B-47 was difficult if not impossible to intercept by contemporary fighters. It was not until the introduction of the century series fighters—and their MiG counterparts—that the B-47 became relatively vulnerable.

The foregoing material might easily lead one to conclude that the B-47 did enough in its military guise to be a classic aircraft, one of the most important of its type. But that is only half the story. The B-47 is also one of the greatest triumphs of capitalism, for it spawned a series of aircraft designs that have made the air transport business one of the most lucrative in history.

Boeing had approximately $10–13 million dollars at risk in the experimental program that led to the XB-47. Had the XB-47 crashed on takeoff and the program been cancelled, the giant Seattle corporation would have been badly hurt, but it might have survived to go on building piston-powered military aircraft for another several years.

But the XB-47 did not crash. It led instead to an increasing number of production orders until more than 2,000 were built. More important, from an economic standpoint, it generated sufficient funds to permit Boeing to create two more prototypes, the KC-135 and the 367-80 (more familiarly known as the 707). The Boeing B-52 would also follow from the B-47 bonanza.

The KC-135 would make history as the first jet aerial tanker. It would revolutionize bomber, fighter, and transport flying, acting as a force multiplier. The United States Air Force could not have conducted a single campaign, from Vietnam to Bosnia, if it did not have its aging fleet of KC-135 tankers, which are slated to serve well into the twenty-first century.

The 707 was even more successful than the KC-135, becoming the first successful jet air liner, and siring the long line of Boeing triumphs which now reaches down to the 777 and beyond. All of these aircraft owe their basic configuration—swept wings and podded engines—to the B-47. And, as Tegler points out, other manufacturers at home and abroad were quick to adapt the B-47/707 formula to their own designs.

The economic payback of the original Boeing $10–13 million investment amounts to hundreds of billions of dollars if one considers aircraft orders alone. If the value of the associated support industry and most importantly the value of air travel itself is imputed to the investment, the return surely reaches into the trillions.

Thus, while the B-47 must (thank goodness) be denied the accolade of being "first in war" it must certainly be given the laurels for being "first in peace," for it precipitated a revolution in commercial air travel that is still going on, and has not yet seen its peak.

The author would like to thank the many B-47 pilots, flight crew, ground crew, Boeing engineers, B-47 Association members, and others whose contributions greatly enhanced this reference including Robert Robbins; George Schairer; Bill Cook, The Boeing Company; George Birdsong; Lloyd Griffin; Bill Slade, 306th Bomb Wing; Don Carey; Robert Hanaway; Richard Ruddy; Charles F. Emmons; John Irving; Jack Wright; Dave Russell; Paul Frye; Leland B. Cook; Raymond Bostoc; Bud East, 40th Bomb Wing Association; Tom Stanton, 509th Bomb Wing; Ben Crouch, 384th Bomb Wing; Mike Cooper-Slipper, Avro Test Pilot; Charles Anderson, USAF & General Electric Test Pilot; Harold "Hal" Austin, 91st Strategic Reconnaissance Wing; Don Griffin; Bruce Bailey; Bruce Olmstead; Jack Kovacs; Jim Nelson; Rolland C. Crane; Dave Johnson; William Bateman; Roby Craft; Robb Hoover; William Henderson, 55th Strategic Reconnaissance Wing Association; Andrew Labosky, 376th BW, 9th SAW; Mark Natola, B-47 Stratojet Association; Robert Burns, NASA; and Charles Koepke, Castle Air Museum. Special thanks also to Walter Boyne; Shelley Carr, The McGraw-Hill Companies; my father, John Tegler, and brother, Eric Tegler.

We had both apprehension and confidence as we prepared for the first flight. I was pretty comfortable, really, because I had a lot of confidence in the Boeing people who were involved in the design: George Martin (XB-47 Program Manager), Ed Wells (Chief Engineer), George Shairer (Boeing's Chief Aerodynamicist), and Bill Cook (XB-47 Head Aerodynamicist) to name a few. Plus, my boss, N. D. Showalter (Boeing Chief of Flight Test), was great. He didn't push us. He gave us a job to do and allowed us to do it. If I'd have wanted to abort at any point, he would have understood and not been critical. That gave us a lot of confidence. The other side of the coin was that it was a truly radical airplane with a lot of unknowns. That first flight of the XB-47 was the only time in my life when, before I released the brakes for takeoff, I prayed for God to please help me through the next two hours. I and many others didn't know what was coming.

ROBERT ROBBINS, Experimental Test Pilot

Just after 2:00 P.M. on the afternoon of December 17, 1947 Robert Robbins sat in the forward cockpit of the XB-47. Lined up on Runway 13 at Boeing Field with the number one prototype's engines at 100 percent, he and fellow test pilot, Scott Osler, were ready to take flight in this incredible airplane. If Robbins and Osler didn't quite know what to expect, they were not alone.

In fact, no one in the XB-47 program knew just how significant and influential the Stratojet would ultimately become. But they did know this new bomber was something special. Viewed through the telescopic lens of history we now recognize Boeing's B-47 as one of the most important aircraft of the twentieth century. It was America's first truly successful jet bomber, the product of cutting edge research and an innovative design which incorporated the most advanced ideas of its day in aerodynamics and propulsion.

The basic configuration of the B-47—its 35-degree swept wing, strut mounted engines, swept horizontal and vertical stabilizers, and aerodynamically clean fuselage—still echoes more than 50 years later. As we enter the millennium, drive out to a local major airport, and scan the ramp, we see variations on a theme. Every modern commercial airliner owes its basic design to the Stratojet. It was the foundation for Boeing's success in the postwar era in both the military and commercial marketplace. Without the B-47 there could be no B-52, no KC-135, no 707. Nor was Boeing alone in recognizing the virtues of its design. Both Convair with the model 880 and Douglass with its DC-8 quickly and successfully followed Boeing's lead.

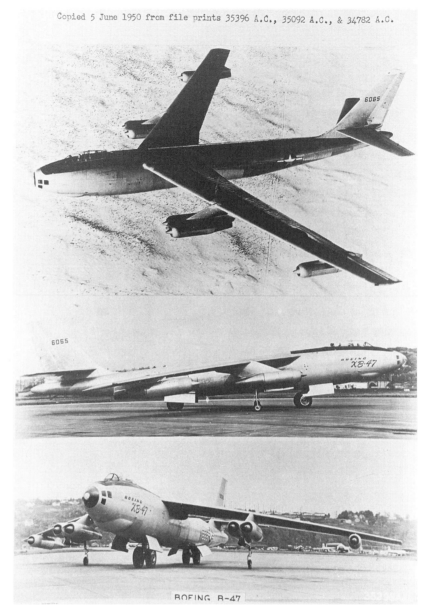

Copied 5 June 1950 from file prints 35396 A.C., 35092 A.C., & 34782 A.C.

Three views of the original XB-47. (*National Archives*)

Of course, the impact of the B-47 extended beyond its technological virtues. With an operational career that began in the early 1950s and closed in the late 1960s, the Stratojet's term of service closely mirrored the tensest, most dangerous years of the Cold War. Before the advent of medium- and long-range guided missile systems, the B-47 was America's chief delivery system for nuclear weapons and arguably our most valuable strategic deterrent.

Innovative and trend setting as it was, the B-47 was not without shortcomings, many of which were present throughout its career. The very experimental nature of its advanced design called for thorough development. The aircraft did undergo a comprehensive flight test program, but a number of factors accelerated the Air Force's desire to make the Stratojet operational, and the aircraft was ushered into production quickly, perhaps prematurely. Still, ask those who flew, navigated, crewed, and maintained the B-47 what they thought of it and you'll get a wide range of heartfelt, largely positive responses. Bob Robbins voices a sentiment often heard from these men: "I think it may very well have prevented World War III."

Such observations were in the future though. On that December afternoon a sense of a drama and tension was palpable among the crowd of engineers, company officials, and family members who trained their gazes toward the far end of Runway 13 watching dirty black clouds of smoke rise from the XB-47's six jet engines. Robbins and Osler were about to discover the airplane's capability.

Origins and Development

As American forces fought their way up the Italian Peninsula in Europe and through the Gilbert Islands in the Pacific during the summer of 1943, USAAF engineers at Wright Field were pondering the possibility of a jet bomber. This advanced thinking was remarkable for a number of reasons. America's first experimental jet fighter, the Bell XP-59, had flown just a few months earlier in October 1942 and was still in early stages of testing. The piston-engined bombers that comprised the Army Air Corps' tactical and strategic bomber force were in heavy production, and the companies that built these aircraft were primarily focused on meeting production schedules, developing existing platforms, and designing heavier, longer-ranged piston-powered bombers. The idea of building a jet engined bomber was foreign to the manufacturers. However, late in the year the Air Corps issued an informal request to five companies, including Boeing, to examine the possibility of designing a multiengined jet bomber based on General Electric's new TG-180 axial flow jet engine.

While Boeing engineers began work on the concept, the Bomber Project Office at Wright Field formulated a menu of performance requirements for the new medium bomber. The preliminary specifications, issued in the spring of 1944, called for a top speed in excess of 500 mph, a service ceiling of 40,000 feet, and a 2,500- to 3,500-mile tactical range. Three manufacturers initially responded to the AAF request: North American, Convair, and Boeing.

Model 424

Boeing's first design study, known as the Model 424, was basically an adaptation of the jet engine to an existing configuration. Featuring a thin, straight, high span wing with engines mounted in twin pods just below each wing several feet from the fuselage, the design resembled a scaled down version of the company's B-29 and was quite similar to the those submitted by North American and Convair. Evaluation of the designs was to take place in NACA's (National Advisory Committee for Aeronautics) wind tunnel. Because the designs were so similar a composite model of the three aircraft was employed for the tests.

Convair's entry in response to the USAAF's requirement for a jet bomber, the XB-46. (*National Archives*)

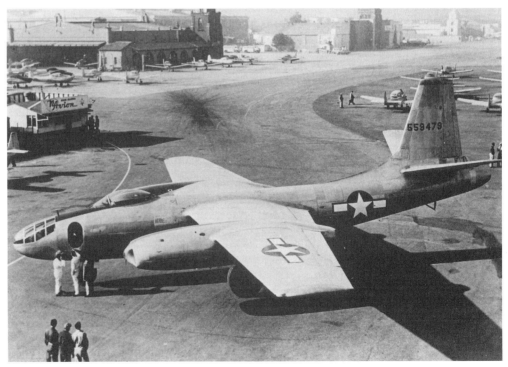

The North American XB-45, one of the four designs competing with the XB-47 for the AAF's jet bomber contract. The B-45 was actually produced and saw service as America's first jet bomber. However, it was quickly dropped in favor of the B-47. (*National Archives*)

The XB-48, Martin Aircraft Company's entry in the jet bomber competition had a similar bicycle landing gear configuration to the XB-47. (*National Archives*)

Meanwhile, an effort was under way that would radically change Boeing's design and the course of aviation history. In late 1944 allied forces were advancing rapidly and it was clear that the war in Europe would end soon. Looking to the future, General Hap Arnold asked famed German aerodynamicist Theodor von Karman to lead a group of the nation's leading scientists to Germany as soon as conditions permitted. "The Scientific Advisory Group" was to look into advanced German research in aerodynamics, radar, rocketry, electronics, and nuclear weapons. Among the experts chosen for this group was Boeing's Chief Aerodynamicist, George Schairer. It was a fortuitous choice. Schairer was assigned to the Pentagon in this capacity in Washington D.C. at the time, far removed from the efforts in Seattle. However, shortly before the group left for Germany, he was advised of an aerodynamic study done by Robert Jones, NACA aerodynamicist. Another member of the Scientific Advisory Group, Dr. H. S. Tsien, told Schairer about Jones's work. Tsien had encountered the frustrated aerodynamicist the day before. Jones had recently authored a report on the benefits of wing sweepback, a concept so seemingly improbable that it was rejected by NACA's editorial committee. Chief among their criticisms was that Jones had no wind tunnel data to bolster his tentative theory. Over the course of the next week Tsien and Schairer discussed the theory and, based upon their own calculations, decided Jones was right. Sweepback could be useful. As the Scientific Advisory Group left for Germany in April 1945, the theory stuck in Schairer's mind.

One of the first sites the group visited was *Reichsmarschall* Goering's Aeronautical Research Institute in Brunswick. In 1939, a young German engineer named Ludwig Bölkow, working for Messerschmidt, had been asked by company officials if a paper on the effects of wing sweepback at supersonic speeds (as advanced by Dr. Adolph Buseman in 1935) might have relevance for the new designs the company was at work on. Bölkow along with Dr. Albert Betz built swept wing models and tested them at the University of Göttingen's wind tunnel. The results of the tests were positive and the studies were published. As a

In this very interesting photo the XB-47 sits alongside one of its competitors, the innovative Northrop XB-49. (*National Archives*)

result work on many different swept wing designs, including the Me-262, was begun by German manufacturers. It was this published work that Schairer found at the Institute. The data amply demonstrated the advantages of sweepback in achieving higher speeds. Further data was uncovered from Junkers and Arado, both companies having been engaged in swept wing designs of their own before VE Day. All of the data seemed to confirm sweepback theory. Armed with the knowledge imparted by Robert Jones, Schairer immediately recognized the significance of the German research. In May 1945 he wrote a letter to Boeing stressing the importance of sweepback, illustrating his point with a calculation based on 29 degrees of wing sweep which clearly demonstrated its potential benefits.

NACA soon completed its wind tunnel testing of the composite model. The results were promising and the AAF decided to award Phase 1 Study contracts to North American, Convair, Boeing, and new entrant in the competition, Martin Aircraft. The four contractors pressed on, refining their designs in pursuit of the performance requirements. Each brought considerable resources to bear but Boeing had an ace in the hole, a wind tunnel. Only three high-speed wind tunnels existed in the United States in 1945: two belonged to

NACA; the third was Boeing's. The company's foresighted construction of a high-speed wind tunnel offered a critical advantage, allowing it to collect data and experiment with new ideas in a way none of the other contenders could.

Upon completion of its own wind tunnel tests of a composite model, Boeing found that the design's drag was unacceptably high, preventing it from achieving the speed called for by the AAF's jet bomber specification. The engine pods mounted on the underside of its wings were considered to be a major source of the drag. Delaying the submission of their prototype design, Boeing engineers searched for a configuration with less drag. Their first step was the Model 432.

Model 432

To eliminate drag and increase wing efficiency the four TG-180s were moved from the wings to the upper area of the fuselage aft of the cockpit above the main fuel tank. Semicircular air intakes fed the engines, which exhausted through tailpipes atop the rear fuselage. It was an alternative solution that had few advantages and didn't solve the real problem. The straight wing itself was the stumbling block, its inherent drag making it impossible to fully exploit the power of the jet engines. Incredibly, the solution was about to arrive in a letter. Sweepback was the answer, the key to achieving higher airspeeds. What had previously been only theory was now confirmed, and George Schairer had the evidence to support it. The theory was straightforward.

In simplest terms, lift is created by a difference in pressure above and below a wing. Air flowing over the upper surface of a wing travels at a higher velocity than an aircraft's actual flight speed. This produces a negative or low pressure area above the wing which results in lift. The airflow on the upper surface of a straight wing begins to approach the speed of sound well before the airplane itself. If the aircraft's flight speed is increased further, the airflow over the upper wing surface will reach the speed of sound and will begin to separate from the aft section of the wing. This separation causes drag and a loss of lift at the least. At worst it can lead to a high-speed stall. This induced "drag rise" prevents a straight winged airplane from achieving higher speeds. The advantage of a swept wing is that it delays the onset of drag rise. Airflow over a swept wing maintains lift at higher aircraft flight speeds because airflow over the wing's upper surface occurs at a lower velocity. This allows the aircraft to travel at higher speeds before the onset of drag or a high-speed stall.

Schairer's letter from Europe made an immediate impact on the XB-47 project engineers who quickly realized the implications sweepback could have for their own design. Aerodynamicist Bill Cook, noted that just one week after Schairer's letter arrived, "the original XB-47 model was cut on its centerline to test a range of sweep angles to verify the general effectiveness of sweep and obtain data on the amount of sweep needed versus the critical airspeed at which drag would rise." Into the wind tunnel went the Boeing engineers to determine a proper sweep angle for the XB-47. Simultaneously North American and Convair were moving ahead with their straight wing designs, making final drawings for construction of their prototypes. With its competitors moving forward Boeing needed to establish a standard sweep angle so that testing of the initial swept wing design could commence. Based on data Boeing had already gathered from wind tunnel tests of various sweep angles, aerodynamicist Vic Ganzer calculated the amount of sweep needed to ensure that airflow over the upper wing surface would not separate at the airspeeds generated by the TG-180s at full thrust. The magic angle was 35 degrees. An inspired bit of engineering in view of the fact that most contemporary airliners employ sweep angles very near 35 degrees and that it was arrived at with little data to go on.

Swept wings were not an entirely new technology, however. Even before World War I swept wings had been employed as a means of shifting the center of wing lift relative to an aircraft's center of gravity (CG) to cure balance problems or permit alternative crew seating. During WWII Curtiss-Wright designed and flew a pusher-powered swept wing fighter, the XP-55 Ascender. In 1945 the newly discovered German research on sweepback was reaching the American aviation community and Boeing was not alone in experimenting with swept wing designs. A sweep angle of 35 degrees was not unique either. Using the German data, Bell Aircraft chose 35 degrees of sweep for the wings of its L-39, a modified P-39 Aircobra which flew in 1946 with test pilot Tex Johnston (who would later fly the XB-47, XB-52, and Boeing model 367, the original 707) at the controls. Finally, North American's famous F-86 Sabre flew just ten weeks before the XB-47. What was unique, though, was Boeing's application of sweepback to a large aircraft. This was an unprecedented and risky undertaking given that swept wings had only previously been mated to small single engine fighter designs.

The challenges facing the XB-47 design team were considerable. Extensive wind tunnel tests of the swept wing configuration had to be executed to prove the feasibility of the design. Moreover, the XB-47 team had to convince the Air Force Bomber Project Office that their radical new design was worthwhile. Doubters within Boeing and the Air Force were many, but to their credit both parties allowed the project to go forward. In an impressively short time the design team arrived at an initial configuration for the swept wing XB-47. Designated the Model 448, Boeing presented the design to the Air Force in September 1945.

Model 448

The Model 448 featured thin flexible wings and horizontal stabilizers swept at 35 degrees and a vertical fin swept at 45 degrees. The fuselage was similar to that of the Model 432 with four engines buried atop the fuselage. Air was ingested through a single intake in the nose section. Significantly, two more engines were added, both fitted to the aft fuselage below the tail, breathing through inlets which straddled the fuselage. The Model 432's range was comparable to that of its competitors (approximately 2,200 miles), but XB-47 program manager, George Martin, presciently predicted that a longer range would be needed if the Stratojet was to come out on top in the jet bomber competition. Calculations indicated that the XB-47 would need the two additional 4,000-pound thrust TG-180s to meet performance and range requirements.

Boeing presented the Model 448 to the Project Office in October. The Air Force rejected the design immediately, citing concerns about the engine location. It was felt that the risk of catastrophic fire due to engine damage was too great in the present configuration. The Air Force directed that the engines be mounted away from the fuselage on the wings in traditional fashion. Thus, another challenge was thrown up before the design team. How would they position the six engines on the thin flexible wing and overcome the drag problems they'd encountered with the Model 424 and 432? They solved the dilemma ingeniously and rapidly. Strut-mounted engines in streamlined, podded nacelles were the answer. Wind tunnel tests proved the logic of the solution. If the nacelles were mounted low enough and forward enough, there would be no negative airflow interference. The wings would have no more drag than they did without the engines. A corresponding advantage of relocating the engines to the wings was that the fuselage could be slimmed, resulting in an even cleaner aerodynamic profile. The Stratojet's configuration had evolved once more. Twin engines hung from pylons situated at approximately one-third of the span. The added TG-180s were housed on single nacelles snugged up directly under the wingtips much in the same manner as the fuel tanks on Lockheed's P-80. The XB-47 was beginning to look like a B-47. Boeing dubbed the new configuration the Model 450.

Model 450

With Air Force approval of the changes, the final iteration of the XB-47 was taking shape. Two major design challenges remained, both related to the wing. Not only was the Stratojet's wing swept, it was very thin. The same concerns over airflow separation and drag which had lead to a swept wing also prompted Boeing aerodynamicists to choose a slender airfoil. This left few options for another important design consideration, the landing gear. A conventional tricycle arrangement was not possible given the thin wing and tightly cowled engines. Some thought was given to incorporating a tricycle gear into the fuselage, but this would have resulted in a significant bulge, spoiling the airplane's clean profile. A further consideration was introduced when the Bomber Project Office, encouraged by the general configuration, added the requirement that the XB-47 be capable of carrying the atomic bomb. In 1945 the atomic bomb was larger than most conventional bombs. Consequently, the XB-47's bomb bay would have to be large enough to accommodate it, limiting available space for landing gear in the fuselage.

Fortunately, the Project Office also proposed an idea to address these obstacles—"bicycle gear." The Air Force had already conducted tests of bicycle gear on a Martin B-26H (nicknamed "The Middle River Stump Jumper," in reference to the location of the Martin plant outside Baltimore) and found it to be a practical solution, allowing for a thin fuselage and the incorporation of a large bomb bay. Boeing took the Air Force's suggestion and began to work out the details of adapting bicycle gear to the XB-47. It was quickly realized, however, that this idea posed a new problem. With the bomb bay located to match the CG of the aircraft the gear would have to be positioned forward of and behind the bomb bay. In fact, the rear gear would be well aft of the aircraft's CG. This arrangement would preclude a traditional takeoff during which the main gear act as a fulcrum as the nose is rotated upward by deflection of the elevators. The elevators would not be able to generate enough force to bring the nose up.

Once more, the XB-47 design team was equal to the challenge. If the aircraft couldn't be rotated into the necessary position for liftoff, they'd put in that position. By building the

A good overhead view of the B-47 planform. (*National Archives*)

aircraft to rest on its gear at the proper attitude, the angle of attack of its wing along with high lift flaps would provide enough lift for takeoff. Aerodynamicist Bill Cook calculated that 8 degrees of angle between the wing chord line and the ground would generate the lift needed for takeoff without rotation. To aid in stabilization during taxi, takeoff and roll out, retracting outrigger gear were built into the inner engine pods. The resulting configuration gave the B-47 a rakish stance. Even sitting still it looked ready to fly.

A final modification of the Model 450 was still to come. As the 1940s wore on it became increasingly clear that the Air Force's postwar philosophy of strategic deterrence favored long-legged, intercontinental bombers. The XB-47 designers still weren't satisfied with the bomber's projected range and began searching for ways to improve fuel efficiency. Realizing that aerodynamic drag could be reduced by increasing wingspan, they decided to extend each wing by 8 feet. The lower drag coefficient resulted in lower fuel consumption and increased range. The extensions were simply added to each wingtip and did not alter the positions of the engines relative to the fuselage.

By early 1946 the basic design of the XB-47 was finalized. Impressed by the speed and creativity with which the Stratojet designers had overcome challenge after challenge and by the airplane itself, the Bomber Project Office negotiated a fixed price $10 million contract for two XB-47 prototypes. The relatively small design team was justly pleased. They had taken their design through several advanced technological evolutions, breaking new ground at every step. Now they could proceed with their experimental project, getting down to the nitty gritty of detail design. As assembly of the prototypes began in June 1946 those involved with the bomber were confident that they were working toward an exceptional new airplane.

Enthusiasm for the XB-47 was not widely shared though. Skepticism about the practicality of such an airplane abounded and many dismissed the Stratojet as a research aircraft which would never be produced. Even inside Boeing there was a relative lack of interest in the project. The company's attention was focused on its current best-seller, the B-50, and

The XB-47 is rolled out for the first time. Strangely, interest in it wasn't very great. As Bob Robbins indicated, only a little over 100 people came to see it. (*National Archives*)

on the B-54, a long-ranged compound-engined bomber based on the B-29 which it hoped would represent its future. As assembly and work on design details such as hydraulic flight controls and assuring longitudinal stability proceeded over the next several months the company's indifference remained. In fact, even as the Stratojet was pushed from its hangar at Boeing Field for the first time—one of the most climactic moments in any aircraft's life— there was almost no fanfare remembers Bob Robbins.

When we rolled the B-47 out in September of 1947 there was very little excitement. If you look at pictures of the rollout you can see that there are very few people around the airplane. If there were 100 I'd be amazed and I think it was closer to 40 or 50. This took place just 300 feet from the headquarters building, 300–400 feet from the engineering building and right out in front of final assembly! That's an indication of the lack of interest in the airplane at the time.

Flight Test

On December 17, 1947, exactly 44 years to the day after the Wright Brothers first trip aloft, a somewhat larger crowd than that present at the rollout gathered to witness the first flight of an aircraft that would have been unimaginable to Orville and Wilbur. Tremendous progress had been made in just four decades, and advanced aircraft made frequent debuts during the period. But nothing like the XB-47 had been seen before. With futuristic looks bordering on science fiction, it made an immediate impression on all who saw it. Certainly, it resembled nothing else on the tarmac at Boeing Field. To start with, it didn't really look like a bomber. Everyone knew bombers were large lumbering beasts. Not the B-47; it was sizable but sleek and anything but lumbering. It's gleaming, streamlined fuselage, swept wing, and tandem cockpit sitting at a jaunty angle on the gear made it more exotic than many of the fighters of the day.

On the ramp, at the far end of the runway, and atop buildings the assembled spectators waited with great expectation for the XB-47 to begin rolling. Relating the tension of the moment aerodynamicist Bob Withington later told Bob Robbins, "Ya know, I went up on top of the wind tunnel to watch that first takeoff. As I looked at the airplane, in spite of all the wind tunnel testing we'd done the thought went through my mind, 'That's a mighty strange looking airplane. I wonder if it really will fly?'" Most of the onlookers shared this thought in one way or another and the drama rose as the engines came to full song in the distance.

Inside the cockpit Robbins and Osler calmly completed their checklist. Though both men felt the natural apprehension associated with any first flight, they did have some idea of what it would be like. Both test pilots had flown the "Middle River Stump Jumper" to gain experience with the takeoff and landing characteristics of bicycle gear. Further simulation was done in Boeing's XB-29 according to Robbins. "Bill Cook did a very good job of telling us how to fly the number-one XB-29, which we still had at Boeing Field to simulate the landing characteristics of the B-47, both the flat approach and the slow engine acceleration. So we had some idea of what to expect."

As it turned out though, things didn't go exactly as planned when Bob Robbins finally released the brakes.

We had a little trouble on the first takeoff run. In taxi tests before the first flight we were able to get up to about 90 mph in the field length that we had and comfortably stop on the runway. I considered 90 mph the refusal speed for takeoff. On the first run, just at 90 mph, a fire warning light came on for the number two engine. I had zero time to make up my mind whether to abort the takeoff or continue. I elected to chop the power. We weren't as far down the runway as we had been in taxi tests and I knew we could stop in that distance. Those were the days when fire detectors in the engines were giving a lot of false alarms. We anticipated the possibility. We had millivolt meters at the copilot station but Scott didn't have time to read them before I had to decide to chop power or go. When he finally did read them they were just on the borderline for turning the warning light on so we knew we had a false alarm. On the actual takeoff, the second try, we had three fire warning lights by the time we were airborne. But Scott was ready for them and he called off the millivoltmeter readings as I yelled, "Fire light on number one! Fire light on number five!" And so on. That gave us a little cause for concern but was really no problem.

The test pilots were airborne in 4,300 feet at 140 mph. "The only other difficulty was that the flaps wouldn't come up initially. So I was faced with either throttling back to stay below flap placard speed or pulling up abruptly to stay slow enough. I elected to pull the airplane up abruptly—which got the attention of the people on the ground—to control the speed and give us a little time to think about it. In spite of all the work that we'd done with them, I didn't really trust jet engines and I swore that I was not going to touch the throttles until I had plenty of altitude unless absolutely necessary. It turned out that there was a just a clutch setting on the flaps that was incorrectly set and by reducing airspeed a little I could bring the flaps up on the emergency system. From that point on, everything was fine. The flight characteristics of the aircraft were good."

Robbins and Osler take off from Moses Lake AFB on one of the first XB-47 test flights. (*National Archives*)

Robbins and Osler make a wide sweep of Moses Lake AFB prior to landing after their highly successful 51-minute, first flight from Seattle. (*National Archives*)

Robbins and Osler leveled off at 15,000 feet at 285 mph and turned east over the Cascade Mountains heading for Moses Lake Airfield. The remote former wartime training base was a perfect site for Phase 1 flight testing. Fifty-two minutes after liftoff Robbins brought the XB-47 in on a long straight in final to Moses Lake. The aircraft came to a stop in 6,000 feet and the first flight was successfully concluded.

Phase 1 Flight Testing

Phase 1 flight testing began in January 1948, and over the next seven months Robbins and Osler logged 43 flight hours on 47 flights in the #1 XB-47. The second prototype made its first flight on July 21, 1948 fitted with new, more powerful 5,200-pound thrust J47-GE-3 engines. These raised the XB-47's performance considerably. Robbins' goal in Phase 1 was simple: "We wanted to give the Air Force an aircraft that was safe to fly and ready for their Phase 2 testing."

Phase 1 flight tests were encouraging. The aircraft performed well and had few negative tendencies in flight. When problems with Dutch Roll (adverse yaw) were encountered in flight they were cured quickly and effectively with Boeing engineer Ed Pfafman's invention of the now ubiquitous Yaw Damper. Early drag tests measuring airspeed provided promising indications of the Stratojet's potential. Bob Robbins had an opportunity to demonstrate its performance to a skeptical observer.

By March of 1948 we were ready to do some good performance tests. Of course, the first thing you need to do good performance testing is a calibrated pitot/static airspeed system. The best way to do that with a big airplane is to have a calibrated pacer plane fly in close formation with you over an instrumented course. The Air Force said, "Fine. We'll send a calibrated pacer up from Muroc [*now Edwards AFB*] to Moses Lake."

They did. It was a P-80 and Chuck Yeager was flying it. Chuck's a hell of a good pilot, but he had a little bit of contempt for bombers and a little disdain for civilian test pilots. Well, we took off, climbed out, and got up somewhere within four or five points of full throttle speed. At that

The number-one XB-47 in flight over the snow-covered terrain of eastern Washington state. (*National Archives*)

point, Chuck called me on the radio and said, "Bob, would you do a 180?" I thought, Hey, Chuck's smart. He just wants to stay reasonably close to Moses Lake. He doesn't have as much fuel as I do. Well, I turned around, got stabilized, and looked for Chuck. He wasn't there. Finally, I got on the radio and said, "Chuck, where are you?" He called back and rather sheepishly said, "I can't keep up with you Bob." So, Chuck Yeager had to admit to a civilian test pilot flying a bomber that he couldn't keep up! That was something!

The clear plexiglass nose of the XB-47. (*National Archives*)

The drag tests were a turning point for the XB-47. The positive performance figures being recorded finally started to generate some enthusiasm with Boeing. Apparently, even as late as midway through Phase 1 tests, the XB-47 had failed to capture the interest of Boeing's top people. Bill Cook reported that he was stunned when George Schairer, the man who'd provided Boeing with the key data on wing sweep, told him, "You can play with your experimental airplane but the bread and butter of Boeing is the B-54."

Phase 2 Flight Tests

The flight test program moved into Phase 2 in July 1948, beginning the Air Force evaluation of the XB-47. Taking on the Air Force flight test card was Major Guy Townsend. Townsend made two major contributions to the program. The first came during flight testing. A significant design defect manifested itself during landings. The Stratojet used up far too much runway. The very cleanliness of design which enabled it to achieve high speeds impaired its ability to slow down in flight and on the ground. The XB-47 had too little drag. Even the common practice of lowering its gear at relatively high speed (maximum gear speed was 305 knots) did little to slow the aircraft or steepen its descent. Consequently, most approaches were long and flat. Once on the runway airspeed bled off slowly, delaying the point at which effective braking could be initiated. Furthermore, the acceleration characteristics of its early jet engines were poor. If a pilot got low and slow on approach he could not instantly call upon the engines for a quick burst of thrust to correct the situation. Power came on too slowly from low engine RPM. Landing fast at higher power was not ideal either as it was very difficult to get the aircraft stopped on the available runway.

Major Townsend made a clever recommendation. Why not slow the aircraft with a parachute? That's just what was done. Boeing engineers installed a ribbon-type drag chute underneath the tail, forward of the tail cone. Deployed just before touchdown, the chute provided the necessary drag to slow the airplane and shorten its rollout. Later, the Air

One of the first test landings of the XB-47 with the new 30-foot-diameter ribbon-type deceleration parachute at Moses Lake AFB. No. 6065 is piloted here by test pilots Scott Osler and James Fraser. (*National Archives*)

Scott Osler and James Fraser, Boeing test pilots, conducting tests of the "ribbon-type" decelera-tion or brake chute. (*National Archives*)

Force applied this inspired improvisation again, adding a second, smaller chute which when deployed in the pattern allowed pilots to maintain a higher power setting. Extra thrust could be commanded quickly because engine rpm was already high. Bill Cook observed that Major Townsend's novel solution may have saved the B-47 program, as attempts to train pilots without the drag chutes would have presented substantial diffi-culties.

Guy Townsend's second contribution also occurred during Phase 2 flight testing. The competition for the jet bomber contract was at a critical stage. North American's XB-45 and Convair's XB-46 had both previously undergone Air Force evaluation. The XB-45 was the more credible of the two designs. Since the XB-47 and Martin XB-48 had not yet flown by the time of the first evaluation, the Air Force decided that it would give the go ahead for production of the B-45. If either the XB-48 or highly experimental XB-47 bettered the B-45's performance, then that aircraft would be put into production and the B-45 would be phased out.

By mid-1948 it was clear that the spectacular performance of the XB-47 far eclipsed that of its competitors. The trouble was, not many people realized just how good it was, espe-cially not the Air Force's decision makers. But that was about to change. General K. B. Wolfe, the Air Force's Head of Bomber Production, headed to Seattle near the end of July to visit with Boeing's founder and president, William Allen. Bob Robbins recalls what happened.

K. B. Wolfe went to Seattle to talk B-50 with Bill Allen and the Boeing people. Bill Allen, smart visionary that he was, said to K. B. Wolfe, "Hey, why don't you stop at Moses Lake on your way back to Wright Field and take a look at the XB-47." Wolfe said, "No, we don't have time. We can't bother with that airplane." Bill Allen pressed him and said, "We'll fly you over in a new B-50 and your B-17 can follow and you'll have time to look at the airplane by the time your B-17 gets there." So they did. Guy Townsend talked Colonel Pete Warden into convincing General Wolfe to go for a ride in the airplane. I was in Seattle at the time, but the story I got was that Guy did

Smoke billowing, the XB-47 makes an ATO takeoff at Wright-Patterson AFB in February 1950. (*National Archives*)

a fantastic job of demonstrating the aircraft to K. B. Ben Werner, who was in the control tower at Moses Lake along with Bill Allen and the other executives, told me that Guy made a pass across the field below the tower and pulled up and the wings just flexed up and up as he climbed. With that acceleration he probably pulled close to the limit [*G*] load on the airplane. Ten days later K. B. Wolfe called on a Saturday or Sunday and wanted to order ten airplanes! It was Guy Townsend that really sold the airplane!

That quickly the Air Force had an all new bomber and Boeing's focus shifted from the XB-54 to the XB-47. Flight testing of the XB-47 went on through 1949 with both Air Force and Boeing pilots including Tex Johnston. Significant problems which were overcome included lateral control at high speeds and high-speed pitch up. The loss of lateral control at high speed was a product of the B-47's thin flexible wings. Aileron deflections at high airspeed caused the wings to twist and bend, so much so that it was feared they could cancel out or reverse aileron action. Spoilers were installed on the upper surfaces of the wing to decrease twist; however, it was determined that this remedy was unnecessary and they were not included on production aircraft.

Sold! The XB-47 sits on the ramp at Moses Lake AFB in July of 1948 just after General K. B. Wolfe's ride with Major Guy Townsend. To the left is a group, including K. B. Wolfe, Boeing president William Allen, Project Engineer George Martin, Guy Townsend, and others. (*National Archives*)

High-speed pitch-up occurred as the aircraft approached the speed of sound. Even with sweepback an outer portion of each wing experienced high-speed stalls as airflow separated from the top of the wing. This often caused the nose of the airplane to pitch up. Pilots had to reduce thrust and push the yoke forward to counteract it. This difficulty was solved with vortex generators: small vanes located in the affected area of the wing. These eliminated airflow separation and pitch-up.

The XB-47 was remarkably successful and so well received by the Air Force that it was accepted for production without any major changes even though upgrades could have been incorporated. These would have taken considerable time and the Air Force was eager to get the Stratojet into service. Thus it was, that on September 3, 1948, the Air Force placed an order for ten B-47A Stratojets for Research, Test, Development and Evaluation (RTD & E) at a cost of $39 million. The Stratojet was on its way.

Performance

Maximum speed   578 mph/502 knots at 15,000 ft;
   545 mph/474 knots at 30,000 ft
Sea level   568 mph/494 knots
Cruise speed   466 mph at 15,000 ft
Stall speed   129 mph at 15,000 ft
Initial climb rate   3,100 ft/min
Service ceiling   41,000 ft
Range   2,650 mi w/a 10,000-lb bombload
Ferry range   4,000 mi

Dimensions

Wingspan   116 ft
Length   107 ft, 6 in
Height   27 ft, 8 in
Wing area   1,428 sq ft

Armament

Two radar directed .50 caliber machine guns
mounted in the tail (not installed on XB-47s)

Bombload

Normal   10,000 lb
Maximum   16, 1,000-lb GP bombs or one
22,000-lb Grand Slam bomb

Crew

Three: pilot, copilot/radio operator/gunner,
navigator/bombardier

Powerplant

Six GE-TG-180s on #1 prototype
later upgraded to J47-GE-3s
on #1 and #2 prototypes

Weights

Empty   74,623 lb
Normal   125,000 lb
Gross   121,080 lb
Maximum takeoff   162,500 lb

Serial Numbers

46-065 and 46-066

SAC's Speeding Bullet

Speed was the requirement which drove the design of the B-47. Speed was the quality which helped sell the aircraft to the Air Force. Speed to launch and climb to altitude quickly, speed to penetrate enemy defenses, and speed to carry the awesome destructive force of the atomic bomb to targets far away. The implications of this need for speed had a bearing on every facet of the Stratojet.

The B-47's introduction to service was called for with utmost speed. Urgent strategic concerns mounted in Washington D.C. and within the Air Force as world events took an ominous turn. First came the Soviet blockade of Berlin and the resulting "Berlin Airlift" in July 1948. Just a few months later President Truman announced that Russia had detonated an atomic bomb. Meanwhile, Communist tensions were rising in the Far East and by June 1950 the United States was involved in a war on the Korean Peninsula. It was through this highly charged atmosphere that Majors Russell E. Schleeh and Joe Howell flew at a record pace from Moses Lake to Andrews AFB, Washington D.C. Averaging 609.8 mph, they completed the 2,290-mile cross country flight in just 3 hours, 47 minutes. They hadn't even intended to set a record remarked Major Schleeh, further explaining that the reason for their speed was simple. The XB-47 made its best fuel economy at high airspeed. Word of the record quickly reached the distinguished group of officials they'd flown back East to demonstrate their new aircraft to in just a few days. After witnessing a maximum performance takeoff and several high-speed passes, their audience—President Truman and several Congressional committee leaders—was suitably impressed. Shortly thereafter, Boeing was informed that the Stratojet would be produced at a rate of 10 to 15 aircraft per month at the government-owned plant in Wichita, Kansas. During WWII Boeing had cranked out B-29s at Wichita. At War's end the facility was shut down but had recently been reopened at the Air Force's request to modernize B-29s and to work on a new project which, if developed successfully and quickly, could drastically improve the mission capability of Strategic Air Command's (SAC) bombers. The Air Force directed Boeing to study air-to-air refueling and to make it practical in a hurry.

Studies began with two modified B-29s, both trailing hoses. When snared by hook and connected, the hoses would allow the transfer of liquid. Then equipment was flown in from

Majors Russell E. Schleeh and Joseph W. Howell land the XB-47 at Andrews AFB, Maryland on February 8, 1949, after completing the 2,289-mile journey from Moses Lake AFB in a record-breaking 3 hours, 46 minutes. Note the whitewall tires. (*National Archives*)

England to test the newly developed British hose and probe method. This procedure was successful but would only allow the transfer of 200 gallons of fuel per minute. The Air Force wanted a transfer rate of 600 gallons per minute. If that goal could be met, the nation's bombers would have the potential to strike any point on the globe. To achieve the desired fuel flow, something larger and more stable than a hose was needed. Boeing came up with a large extendible flying boom which when directed or "flown" into a trailing aircraft's refueling receptacle by a "boom operator" could transfer the desired amount of fuel quickly. It was in Wichita that this revolutionary capability was developed and in Wichita that the B-47 would be built. The two were intertwined. Aerial refueling was the Stratojet's enabler, something the short legged B-47 couldn't do without. The Stratojet's wings were so thin that there was no room for traditional wing fuel tanks. (During the last XB-47 wind tunnel tests it was discovered that an airfoil need not necessarily be uniformly thin to reduce drag. Airfoil thickness could have been varied. This would have left room for wing fuel tanks, but the wing had already been built by the time of this discovery so no change was possible.) Thus, all of the B-47's internal fuel had to be stored in the fuselage. Seventeen thousand gallons were distributed in tanks forward and above the weapons bay, and behind it in the aft fuselage section. In-flight fuel consumption had to be monitored carefully to maintain the aircraft's CG.

> The CG was important to critical. Managing it, per se, was not difficult. Like most everything else about the aircraft, one had to pay attention. The fuel control panel, statistically, had 128 possible combinations of settings. Some, if not most of those, were not of practical significance, however possible. But this shows the degree of complication and how easy it was to mess it up. Some variants had different CG limits as I recall. It was probably one of those things that anyone who ever flew the beast mismanaged sometime. Whether you got bit was the question. A couple of times I did but was able to correct the situation.
> ROLAND "CRANE" SMILEY, RB-47E/H, ERB-47H, EB-47E PILOT, 55TH SRW, 1960–1965

Early jet engines gulped prodigious quantities of fuel. Even with 17,000 gallons of internal fuel the Stratojet's range didn't satisfy the Air Force, especially SAC. General Curtis LeMay, the commander of SAC, was adamant. The country's new strategic bombers must have intercontinental range. With a maximum mission range of 2,650 miles the Stratojet was not in that class. When the XB-47 rolled out in September 1947 General LeMay was less than enthusiastic. "We have no requirement for production of the airplane. You have to get some range in it."

Aerial refueling was the solution to the range issue. With the development of boom-type refueling Boeing itself solved the problem. As long as the B-47 could get fuel in flight it could reach any target. This advance made the Stratojet much more attractive to the Air Force. Boeing further aided its cause with the inspired adaptation of its C-97 Stratofreighter to the KC-97 aerial tanker configuration. Henceforth, wherever there were Stratojets, there were KC-97s. Now, the B-47 could stand alert, whether deployed overseas or sitting on a ramp in the Midwest and deliver the same punch.

Getting the B-47 into the air to deliver that punch could at times be challenging. The B-47 had to literally be flown off the runway at the 8-degree angle of attack its gear gave it. Not surprisingly, takeoff runs were long. Even the very lightly loaded XB-47 used up

The second XB-47 makes a rocket-assisted takeoff at Moses Lake AFB in November 1948. (*National Archives*)

4,300 feet of runway on its first flight. Production models loaded with fuel and weapons could weigh in excess of 200,000 pounds. On a hot summer day at a high gross weight the B-47 needed every bit of available runway to get airborne (especially at overseas bases where runways could be short). Measures taken to help get the Stratojet off the ground more quickly initially included thrust augmentation in the form of jet assisted takeoff units (JATO).

JATO, or ATO units as they came to be known, were integrated into the fuselage of the XB-47, B-47A, B-47B, and early B-47E models. Nine bottles were positioned on either side of the fuselage in rows of three, above and adjacent to the rear main wheel well. The early 18 bottle units generated 1,000 pounds of thrust apiece, reaching 80 percent of maximum thrust within a half-second after ignition. Once ignited, the internal ATO units launched the Stratojet into the air in impressive fashion, leaving a trail of billowing smoke. Normally, ATO systems were ignited at a precalculated point during the takeoff run, several seconds prior to rotation. Burning for approximately 15 seconds, the effect was noticeable, shortening takeoffs and boosting initial climb rates substantially. Later, more powerful ATO units were fitted to the Stratojet.

> **Two types of external ATO racks replaced the internal ATO. The mounting was just aft of the rear main wheel well on the bottom of the fuselage. The first type of rack was called a *horse collar rack* and held 33 bottles of solid rocket propellant, configured in three rows of 11 bottles. The second type of rack was called a *split rack* and held 30 bottles with three rows of 5 bottles mounted on each side in a V configuration.**
>
> **It was around 1958 at Schilling AFB, Kansas, that I started making ATO takeoffs. We had an annual requirement to make one of these takeoffs. We just read the pilot's handbook about ATO and checked ourselves out. We went into the performance charts and computed an ATO fire speed and lifted a red guarded switch on the number one throttle. The solid propellant was fired electrically and could not be stopped once the switch was activated. The burn time was about 15 seconds. The ATO fire speed was designed to give the pilot seven seconds of acceleration on the runway before liftoff and 8 seconds of acceleration in the air.**
>
> **There was a noticeable increase in acceleration when you lit the fire and a noticeable decrease when the fuel burned out. It was not a rocket ride to the moon. After takeoff, we would fly over the Camp Phillips Bomb Range at about 1,500 feet and release the rack. (*This was a standard procedure fleetwide—racks were jettisoned in designated drop zones.*) This was accomplished by pulling up on a T-handle in the pilot's cockpit. The racks were for one-time use.**
>
> **One of the problems we encountered early on was a CG problem at liftoff. The centerline of thrust of the 33 bottles was ahead of the CG of the airplane. This caused the nose of the aircraft to rise dramatically, sometimes too much. This happened to a pilot I knew. He told me that if the ATO had not burned out when it did, he would have stalled, even though he had full forward control column pressure. Armed with this knowledge, I burned 3,000 pounds of fuel out of the aft main tank prior to my ATO takeoffs. This worked quite well and I did not experience any control problems.**
>
> **RICHARD RUDDY, B-47E PILOT, 44TH BS, 40TH BW, 1954–1964**

The use of internal and external ATO units seems to have been somewhat limited, however. Different bomb wings had different policies, some requiring their aircrew to make annual ATO takeoffs, others rarely if ever using solid rocket propellant thrust augmentation. ATO was used primarily by aircraft standing alert or in emergency circumstances such as the issuance of SAC's Emergency War Order. When transmitted this order would have entailed getting B-47s off the ground in a hurry, no matter the gross weight of the aircraft or available

The XB-47 shows off its clean lines over eastern Washington in 1948, a year during which Phase I and II flight testing were in progress. (*National Archives*)

The first XB-47 on the ramp at Andrews AFB, Maryland after its record-breaking 3-hour, 46-minute flight from Moses Lake AFB. Note the no. 3 engine bullet. It was lost during the approach to Andrews. (*National Archives*)

An early B-47E conducts maximum takeoff load tests at Edwards AFB using the new external, jettisonable "horse collar" ATO rack. An F-80 flies chase. (*National Archives*)

Out with the old and in with the new. The XB-47 and the aircraft it would replace, the Boeing B-29 Superfortress. The two aircraft are approximately the same size, yet the B-47 had a crew of only three. (*National Archives*)

A KC-97 copilot gives his crew chief the "OK" sign after completing a preflight. The B-47Es in the background at Lake Charles AFB would not have been credible strategic bombers without the aerial refueling provided by each Wing's KC-97 squadron. (*National Archives*)

runway. ATO was fairly reliable but not without dangers. Bottles could break free from their mounts and ignite nearby fuel tanks or combust vapors exhausting from fuselage fuel vents with predictable results. Finally, the units were expensive, another factor which limited their use and, as we shall see, engine development lead to other types of thrust augmentation.

Speed was a factor in two more design elements of the Stratojet. Since their invention bombers had been growing in size, both to increase their range and payload and to bolster their defensive capability. Accordingly, bomber crews had also grown in size. The B-17 and B-29 flew with a 10-man crew, the B-50 with a crew of 12. But the B-47, an aircraft of similar proportion to the B-29, performing essentially the same mission, featured a crew of only three—pilot, copilot, and navigator/bombardier. The three crew members shared a pressurized compartment forward of the wing and weapons bay. This compartment was a marked departure from past configurations. The clean, fast fuselage was topped by a two-place plexiglass bubble canopy which housed the pilot and copilot. The tandem cockpit was much more like those found on contemporary jet fighters and trainers than the flight deck of a plodding bomber. Offering excellent visibility and a much "sportier" feel it was another by-product of the aircraft's aerodynamic quest for speed. Pilot and copilot opinions vary about the relative efficiency of the configuration and smaller crew size, but most agree that operating the Stratojet demanded close attention from each crew member.

> The crew of a B-47 bomber was the busiest in the business. It was a big bomber that required a lot of attention and there were just three crew members to do all the work. I also flew the B-52G for several years as a crew commander out of Griffiss, AFB, and although it was bigger with more engines than the B-47, I always felt it was a much easier aircraft to fly. It had a roomier, side-by-side cockpit and six crew members to share the work.
> **CHARLES F. EMMONS, B-47B/E PILOT, 44TH & 45TH BS, 40TH BW, 1958–1964**

The navigator's station was more conventional, situated in the nose of the Stratojet. The earliest models featured clear plexiglass noses, while later versions had only small plexiglass panels on either side of the nose or none at all. Generally, the navigator's station was satisfactory but it did have a few drawbacks.

Missions were flown at all hours of the day in every climate you can imagine; however, the most uncomfortable were the daylight missions. The reason for this was that the pilots sat under the canopy, which gave them extra heat due to the greenhouse effect while the navigator sat in the nose freezing. In fact, the nav's floor was so cold that anything left on it would freeze. Later, an attempt was made to improve conditions by adding an auxiliary heater in the air vent by the navigator's feet. It didn't work very well but it was better than nothing.

JIM "FAROUK" NELSON RB-47K/H/TT NAVIGATOR, 55TH SRW, 1958–1964

Finally, speed influenced the defensive armament of the B-47. When the Stratojet first flew in the late 1940s its performance exceeded that of just about anything else in the sky. With a top speed of nearly 600 mph and a ceiling approaching 40,000 feet, not even early jet fighters could stay with it. In addition, Boeing's experience with stripped, high-speed B-29Bs during WWII suggested that bombers capable of scorching along at fighterlike speeds could only be successfully attacked from the rear. These factors, along with aerodynamic considerations, led to the provision for a sole, unmanned twin turret mounted in the tail of the aircraft. Initially featuring .50 caliber machine guns, the turret could be operated manually or automatically by the copilot. In manual mode the copilot could aim and fire the .50s by employing a radar scope to aid in gun laying. In automatic mode a fire control system (FCS) detected and tracked targets. An onboard computer calculated ballistic equations which positioned the guns and electrically fired them at the proper moment. The reliability and accuracy of the early FCS were less than optimal. An improved FCS and an upgrade to 20-mm machine guns did improve capability and reliability but problems still arose. Stratojet bomber crews were seldom required to fire their guns, perhaps once a year for credit. But the Reconnaissance/ELINT Stratojet crews of the 55th Strategic Reconnaissance Wing fired their guns on every training mission. With good reason. The recon/ELINT missions flown by 55th crews often put them in harm's way. Flying just off and sometimes into Soviet airspace to gather intelligence on Russian defense systems frequently brought them into contact with Soviet fighters. Bruce Olmstead, the copilot of a 55th RB-47H which was shot down by Mig-19s over the Barents Sea is one of the few men who had an opportunity to use the B-47's defensive armament in combat.

Whenever it was possible, on nearly every training mission, we fired our guns. We normally did that over Matamoros, Texas. What we'd do was dump a bundle of chaff out the back of the airplane and then lock onto it and fire at it. With the manual system you had to designate the target. I had to go for the bug, a bright dot that would show up on the radar screen when you locked onto a target. The radar dish, which was mounted well aft in the fuselage, was controlled by a handle on the right side of the cockpit. As you moved the handle around, the radar searched for a target and a horizontal line swept across radar screen. When the radar found a target, the line would sweep across the target and the bug would appear. The handle had a button on the inboard side of it. You'd move it up and down or left and right and when you put a cursor on the target and let go of the bug the radar would lock on. Then the guns would track the target and you could fire. I shot down two drones in gunnery school at Schilling AFB which qualified me. When we dumped the chaff I'd manu-

A crew sit poised and ready in their B-47E at Pinecastle AFB. The crest on the right side of the nose denotes the bomb wing to which the aircraft belonged, in this case either the 19thBW or the 321st BW. (*National Archives*)

ally lock on it with the handle and fire. After we landed we always worried about some of those rounds cooking off if we hadn't emptied the guns.

But when I was confronted with a MiG, a real airplane, my scope was absolutely white. I think he had a buzzer [jammer]. I had the gun radar on all the time, and when we saw this guy comin' down the pike I looked at my screen and it was just white.

BRUCE OLMSTEAD, COPILOT 343RD SRS, 55TH SRW, 1959–1961

Like every aircraft which had come before it and those which followed, the B-47 was a compromise, a trade-off which sacrificed certain traits in pursuit of a defined goal—speed. Empowered by aerial refueling, the Stratojet was a lightning quick intercontinental bomber that could hit hard. The nation got a new strategic deterrent as SAC took delivery of its speeding bullet!

The front office. The pilot's cockpit of the Stratojet. (*National Archives*)

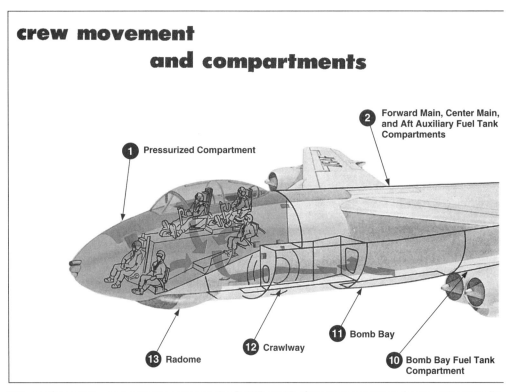

This configuration was much the same in the B-47B and E. Crew movement and compartments diagram from a B-47E Dash-One flight manual. (*Jack Wright*)

Another view of the pilot/aircraft commander's cockpit. A battery of dials, gauges, switches, and buttons faced the B-47 pilot. Here, the pilot's hand holds the drag chute release. Also to hand are controls for air-conditioning, engine starters, ATO, anti-icing, bomb doors, steering, the gear handle, landing and taxi lights, and the battery. Note the large trim wheel below the pilot's arm. (*National Archives*)

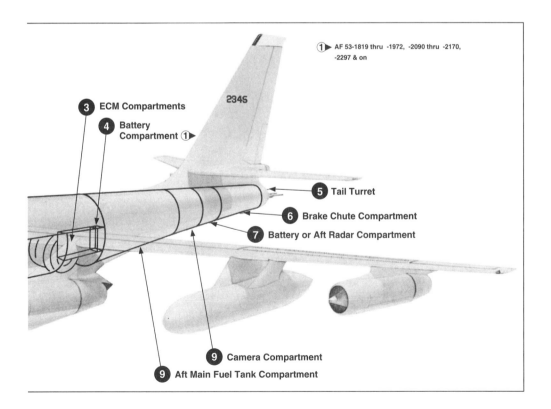

B-47A

The morning of June 25, 1950 was an eventful one, laden with significance for the Stratojet. In Kansas, the first B-47A off the Wichita production line made its maiden flight. Six thousand, five hundred thirty-one miles away the North Korean People's Army was streaming over the 38th Parallel with seven assault infantry divisions, a tank brigade and two independent infantry regiments. The Korean War had begun and its impact upon all military aircraft, including the B-47, would be felt quickly. However, the A model Stratojet, ordered by the Air Force in September 1948, would never really become a combat ready aircraft.

Ten B-47As were built at Wichita over the course of the next year. Their configuration was largely unchanged from the XB-47 prototypes. The takeoff weight of the B-47A went up approximately 30,000 pounds to 151,324 pounds from the XB-47's 121,080-pound takeoff weight. This began a trend in weight gains that would continue throughout the life of the Stratojet, often zeroing out the benefit from successive additions of more powerful, updated versions of its GE engines. The B-47A carried six J47-GE-11 turbojets providing the same 5,200 pounds of thrust as the J47-GE-3s fitted to the second XB-47 prototype and retrofitted to the number one XB-47. Thrust augmentation was identical to that used in the XB-47 with the same internal 18-unit ATO system.

Of the ten A models, only two had defensive armament. Boeing and the Air Force were evaluating two fire control systems for the Stratojet's tail-mounted .50 caliber machine guns. The Emerson A-2 FCS was installed on the seventh A model, while the ninth was tested with an early version of GE's A-5 FCS. Offensive armament systems were installed in four of the A models. The K-2 bombing and navigation system, consisting of an HD-21D autopilot, a computer, the APS-23 radar, and the Y-4 or Y-4A bombsight, was employed to drop the B-47's load of conventional and atomic bombs.

The K-2 was said to be unreliable and it had its drawbacks but let's remember, it was a brand new piece of equipment, kind of a leftover from B-29/B-50/B-36 systems. It was old technology. Some of that equipment was built for use at 15,000 feet, and here we were

A small crowd gathers at Boeing-Wichita as the first B-47A is rolled out. (*National Archives*)

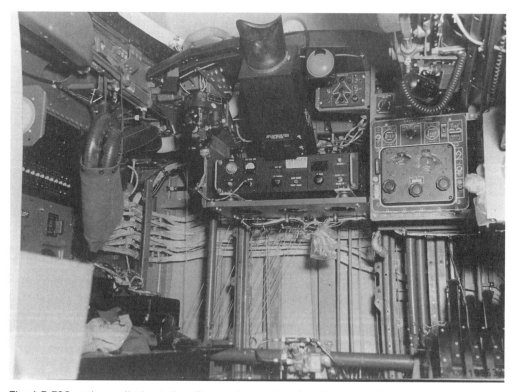

The A-5 FCS at the copilot's station. The copilot could swivel his seat to face rearward to operate the FCS and 20-millimeter cannon. At the center is the radar scope, the turret control, and IFF. On the left panel are ejection seat arming buttons and the chaff dispenser buttons. (*National Archives*)

The first B-47A on a test flight from Boeing-Wichita. (*National Archives*)

flying at 40,000 feet trying to see the ground with it. Neither the radar nor the antenna were good enough. Also, a lot of people that were flying the first radar sets and bomb-navigation systems came from other airplanes or weren't used to radar. There were other factors as well. You could be called on the carpet if you had a bad bomb [*a poor performance on a bomb run*]. So many times, navigators would come back from a mission saying, "Well, it's the equipment's fault. I didn't do anything wrong." Later systems were better but I always wondered if half of the maintenance malfunctions which were reported and were listed as "could not duplicate" just gave the system a bad name because at that time the art of bombing by radar was still being perfected. It had been in development since WWII but was now being done at a higher altitude and twice the speed, which made it more difficult. That's kind of a pitfall in trying to judge what was good or bad about it. It was only as good as the training of the operator, maintenance, and the state of the art.

ANDREW LABOSKY B-47E/EB-47E NAVIGATOR-BOMBARDIER 1960–1966

Conventional and nuclear bombs were dropped from the aircraft from a long bomb bay identical to that of the XB-47. The B-47A also featured the same plexiglass nose and round-tipped vertical stabilizer as the XB-47. Ejection seats were present in the A model Stratojets. Both pilot and copilot ejected upward while navigators ejected downward, exiting the aircraft through an escape hatch on the bottom of the forward fuselage below the nose. At high altitude this was no problem, but at low altitude or on takeoff navigators were at a disadvantage. If the aircraft was too low they stood little chance of survival in the event of ejection. Safe ejection altitude, depending on airspeed, was about 500 feet above the terrain in level flight.

B-47As were mainly used as service test aircraft and for aircrew transition training by the Air Force, and for further flight test by Boeing. The company began deliveries of the B-47As to the USAF in December 1950. There was a great deal of excitement about the B-47 within the Air Force, and a fierce competition broke out among SAC's bomb wings to be the first to get the Stratojet. Ultimately, the honor went to the 306th BW based at MacDill, AFB, Florida. Their transition from B-29s and B-50s to the Stratojet began in May 1951. The

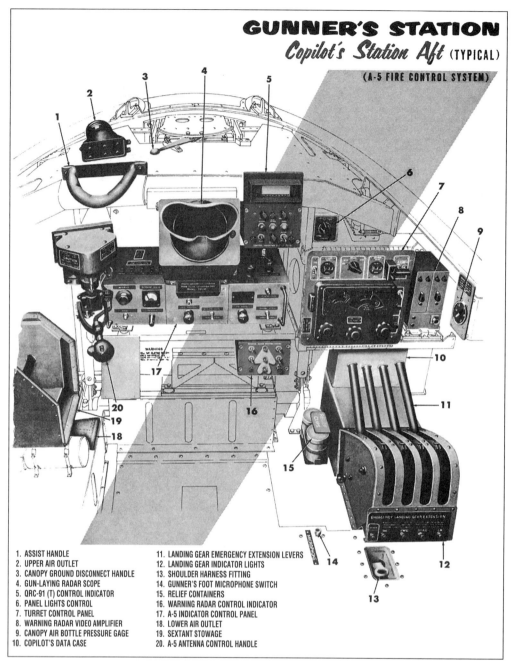

GUNNER'S STATION
Copilot's Station Aft (TYPICAL)

(A-5 FIRE CONTROL SYSTEM)

1. ASSIST HANDLE
2. UPPER AIR OUTLET
3. CANOPY GROUND DISCONNECT HANDLE
4. GUN-LAYING RADAR SCOPE
5. QRC-91 (T) CONTROL INDICATOR
6. PANEL LIGHTS CONTROL
7. TURRET CONTROL PANEL
8. WARNING RADAR VIDEO AMPLIFIER
9. CANOPY AIR BOTTLE PRESSURE GAGE
10. COPILOT'S DATA CASE
11. LANDING GEAR EMERGENCY EXTENSION LEVERS
12. LANDING GEAR INDICATOR LIGHTS
13. SHOULDER HARNESS FITTING
14. GUNNER'S FOOT MICROPHONE SWITCH
15. RELIEF CONTAINERS
16. WARNING RADAR CONTROL INDICATOR
17. A-5 INDICATOR CONTROL PANEL
18. LOWER AIR OUTLET
19. SEXTANT STOWAGE
20. A-5 ANTENNA CONTROL HANDLE

The copilot's aft gunner's station found on all B/RB-47s. The copilot could swivel his seat 180 degrees to operate the A-5 FCS and guns. (*Dave Johnson*)

306th was originally earmarked to instruct future crews transitioning to the B-47, but its role grew quickly. The unit was given the task of developing concepts and techniques for employing the B-47 and making it combat ready. The Wing essentially wrote the operational manuals for the Stratojet and proved its combat capability over a 2-year period which culminated with the aircraft's first deployment.

Not all of the ten B-47As went to the 306th BW. Some remained in Wichita for further company experimental testing, including in-flight refueling tests. Though the A models were phased out of service during late 1951/early 1952, one B-47A did continue logging productive flight time for several more years. The first B-47A built (49-1900) was obtained

The B-47A displays its long bomb bay doors and larger bomb bay which could accommodate the large early nuclear weapons. (*National Archives*)

The first B-47A on a test flight from Boeing-Wichita. Note the clear plexiglass nose and rounded tip of the vertical stabilizer. (*National Archives*)

The first B-47A built (49-1900) was obtained by NACA's High Speed Flight Research Station (HSFRS) in 1953 to study the characteristics of large, flexible swept wing aircraft. Here, the aircraft sits on the ramp at Edwards AFB, CA. (*Robert Burns*)

by NACA's High Speed Flight Research Station (HSFRS) in 1953 to study the characteristics of large, flexible swept wing aircraft. 49-1900 became NACA 150 and was flown from the NACA HSFRS at Edwards AFB from May 1953 to 1957. Studies of aeroelasticity resulted in reports which provided information to engineers and design teams around the country on the behavior and characteristics of large swept wing aircraft like the KC-135 and B-707.

B-47A 49-1900 became NACA 150 and was flown from the NACA HSFRS at Edwards AFB from May 1953 to 1957. Studies of aeroelasticity resulted in reports which provided information to engineers and design teams around the country on the behavior and characteristics of large swept wing aircraft like the KC-135 and B-707. (*Robert Burns*)

A view from below the first B-47A. (*National Archives*)

Contemporaries: The Boeing B-47A and the North American F-86 Sabre. Both benefited from the research on 35 degree swept wings. Seen here in 1951, the pair were, respectively, the fastest bomber and fighter in the world. (*National Archives*)

Even as the B-47A entered service in May 1951 B-47Bs had begun to roll off the Wichita production line. In just five short months the 306th BW would take delivery of the first true production model of the Stratojet, the B-47B.

Specifications for the B-47A are as follows:

Performance

Maximum speed   600 mph/521 knots
   at 8,800 ft
Service ceiling   38,000 ft
Combat ceiling   44,300 ft
Initial climb rate   3,375 ft/min
Range   2,650 mi w/10,000-lb bomb load
Combat radius   1,550 mi
Ferry range   4,000 mi

Dimensions

Wingspan   116 ft
Length   106 ft, 9 in
Height   27 ft, 8 in
Wing area   1428 sq ft

Armament

Two radar-directed .50 caliber machine
   guns mounted in a tail turret

Powerplant

Six J47-GE-11 axial flow turbojets of
   5,200-lb thrust each

Bombload

Normal   10,000 lb
Maximum   16 1,000-lb GP bombs or one
   22,000-lb Grand Slam bomb

Weights

Empty   73,240 lb
Normal   106,060 lb
Maximum   151,324 lb

Crew

Three: pilot, copilot/radio operator/
   gunner, navigator/bombardier

Serial Numbers

49-1900 through 49-1909

B-47B

It was September 1951 and I was in 10-25 school
when I first saw the B-47. We flew up to Wichita and
went out to the Boeing factory. We looked at the production line and then went into a hangar at the end of
the line and looked at one that had just been finished.
In a pilot's words it was, "the airplane." It was large
with those six engines and the wings were drooping
down which looked odd. But it reminded me of sheer
speed. Sitting still, it looked like it was going 100 mph.

GEORGE "BUZZ" BIRDSONG, PILOT/369TH BS C.O.,

306TH BW 1952–1954

We had watched the B-47 being built at Wichita. When
I first saw it I thought, "I wanna fly it!" The first one I
saw had white sidewalls on the tires!

LLOYD GRIFFIN, PILOT/367TH BS C.O.,

306TH BW 1952–1956

When I first saw the B-47 I thought, "Boy, that's a big
mother!" I saw the drooping wings and six engines and
said, "That's gonna be a honey to fly!" The B-47 just
grabbed all of us when we first saw it because we'd
never seen anything like it before.

BILL SLADE, PILOT, 369TH BS, 306TH BW 1952–1956

The reactions of these three men were typical of those voiced by new crew members
upon first seeing the Stratojet. George Birdsong, Lloyd Griffin, and Bill Slade were
among the first operational pilots of the B-47, and each would be a part of the 306th
Bomb Wing's pioneering role in B-47 operations. The airplane they would transition to was
the first truly operational production model of the Stratojet, the B-47B.

Eighty-seven B-47Bs were ordered by the Air Force in November 1949. Less than two
years later the first example flew on April 26, 1951. Lessons learned during the XB-47/
B-47A test program lead to a number of changes and improvements to the B-47 which were
evident internally and externally. Looking over the aircraft from stem to stern, one first

The B-47B featured new 1,780-gallon external fuel tanks, more powerful J47-GE-23 engines, and twin .50 caliber tail guns, which are absent from this B-47B photographed in March 1952. (*National Archives*)

noticed the revision of the nose section. The plexiglass nose cone of the XB-47/B-47A was replaced with a solid cone, featuring four small glazed windows on the left side and two on the right side. The change was due in part to the incorporation of a single point air-to-air refueling receptacle mounted atop the nose and offset to starboard. This modification to allow in flight refueling was critical to the aircraft's success.

Scanning further down the fuselage behind the forward main gear, a shorter bomb bay was immediately noticeable. The miniaturization of nuclear weapons permitted a reduction in the size of this compartment which, nevertheless, could carry 18,000 pounds of nuclear or conventional bombs—a bombload larger by 8,000 pounds than that carried by the B-47A. Out on the wings an effort to augment the aircraft's range was manifest in the form of two auxiliary fuel tanks. Mounted on pylons between the inboard and outboard engines, the jettisonable external tanks carried 1,780 gallons of fuel each. Finally, a change to the empenage was visible. The vertical stabilizer was revised with the squaring off of the fin tip.

Changes inside the B model centered on electronics and weapon system upgrades. The K-4A Bomb/Nav system was installed in the majority of B-47Bs (early models retained the K-2 system), featuring a periscopic bomb sight (visible on the nose), an AN/APS- 54 warning radar, and an AN/APT-5A electronic countermeasures system. These were integrated with mechanical mission computers and an autopilot. Defensive armament consisted of twin-mounted .50 caliber machine guns firing 600 rounds per minute. The guns were controlled by the early B-4 FCS and were aimed using the system's radar sight. The radar sight was not altogether reliable and some B-47Bs were retrofitted with the N-6 optical sight. Structural modifications considerably strengthened the Stratojet, allowing it to carry heavier fuel and weapons loads. Fully outfitted with mission equipment and external fuel tanks, the B-47B could now take to the runway with a takeoff weight as high as 200,000 pounds. Mindful of the aircraft's weight gains and marginal power, the Air Force made a rather poorly considered request.

The tandem pilot/copilot positions in a Stratojet were unlike the configuration in any previous large bomber. Its large plexiglass canopy offered excellent visibility and gave pilots the feel of flying a fighter rather than a bomber. (*National Archives*)

A rear-view of a B-47B, showing its slim profile, Fowler split flaps, and tail cone with twin .50 caliber machine guns. (*National Archives*)

The twin .50 caliber tail guns of a B-47B. (*National Archives*)

George Martin, the B-47 program manager, came into my office one day in the latter part of 1949 with a letter from the Air Force that directed Boeing to remove the ejection seats from the B-47B production aircraft. The reason was to save weight. George said to me, "Bob, put that letter in a safe place. The day is going to come when the Air Force is going to regret this decision, and we want to be able to make it very clear where the decision came from to do away with the ejection seats."

ROBERT ROBBINS, BOEING EXPERIMENTAL TEST PILOT

Escape in the event of an inflight emergency now meant using the crew entry door at the bottom of the fuselage for egress. A spoiler underneath the nose was fitted to make bailing out safer. But it was of little aid. Crew members wearing full flight clothing and parachutes would have had a difficult time escaping the Stratojet, even in controlled level flight. If the aircraft departed level flight and went out of control, escape would have been all but impossible. The decision was clearly a mistake, a matter of being pennywise and pound

Capt. Edward G. Sperry was among a group of volunteers to test the B-47 navigator ejection seat. The Wright Air Development Center conducted four successful ejections with volunteers. Unlike the pilot and copilot, the navigator ejected downward, firing through an escape hatch on the bottom of the nose. Here, Sperry ejects through the hatch and rides the seat downwards. (*National Archives*)

foolish. Deletion of the ejection seats saved a small amount of weight but certainly not enough to offset the aircraft's weight gains.

Crew members could not have been overjoyed to learn that the safest, most effective means of emergency egress was not available in the B-47B, but they were not deterred. It was a fact of life at the time, and the young crewmen who would get their hands on SAC's new bomber didn't raise a ruckus about it as some sources have suggested.

The pilot and copilot of a Stratojet ejected up and out in the normal fashion. Here, a Boeing B-47 ejection seat test sled ejects a dummy on a test run at Edwards AFB, Ca. (*National Archives*)

B-47 ejection seat testing for the navigator was undertaken in the early 1950s. After directing that ejection seats be omitted from the B-47B in the interest of saving weight, the Air Force realized its mistake and reversed its decision, restoring ejection seats in the B-47E and retro-fitting many B-47Bs with them. In this photo, a volunteer is successfully ejected at 500 mph at 10,000 feet. (*National Archives*)

Even without ejection seats I can't think of anybody in the outfit (306th BW) who was unhappy about it. We were just so damned pleased to be flying the B-47 we didn't care what they had. No one ever refused to fly. Everyone was so eager to get into the aircraft.

GEORGE BIRDSONG

Interest in the B-47 was indeed high. In the ready rooms of SAC bases around the country pilots accustomed to flying several generations of piston-powered bombers were looking forward to strapping on the technically advanced Stratojet. The first Air Force crews to transition to the B-47 were a talented and skillful group, many of whom had wartime experience and thousands of hours in their logbooks. They were selected because of that experience and they were well aware of the potential dangers inherent in pioneering operations in a new aircraft, especially an early jet. This did not dampen their enthusiasm however.

The B-47 was a fundamentally safe aircraft if flown correctly. If the aircraft was not flown correctly or was the victim of an engine or airframe failure, the consequences could be terrible. Here, a base photographer photographs the wreckage of a B-47E that crashed on takeoff from Barksdale AFB. (*National Archives*)

We were extremely happy to get an airplane with some real speed. I'd been flying B-17s and B-29s, and in those B-29 units there was a great deal of excitement. We were looking forward to stepping into a new, sleek jet-powered airplane, something that could really move!

GEORGE BIRDSONG

I had flown B-17s, B-24s and B-50s, and, of course we were excited about the B-47. Everyone I knew wanted to get into it. I can't think of anybody that didn't want to be in that airplane.

LLOYD GRIFFIN

The small initial cadre of 306th BW pilots who flew the few B-47As which briefly served with the Wing received their transition training in Wichita at the Boeing facility. Their train-

A mix of B-47Bs and early B-47Es. One wonders what the phrase "To Proton" on the B-47B in the foreground meant. (*National Archives*)

ing focused soley on flying the aircraft. But the training program for the first crop of crewmen who would fill the 306th's three squadrons and help make the Stratojet operational was much more involved and ambitious.

Classes were started in Air Training Command specifically for B-47 crews. The Air Force wanted to put the crews through a triple ratings system, which we called *The Three Headed Monster.* The idea came from General LeMay who'd flown as a navigator. He wanted us to have three ratings. We had to qualify as a pilot, a bombardier/navigator, and a radar operator. That way any crew member could perform any one of the jobs if necessary. It was called *10-25 school,* and once you'd been through it you were 10-25 qualified.

GEORGE BIRDSONG

We all had to go through school. We started in Houston at Ellington Field for navigation and radar training. Then we went out to Mather Field, California for bomb/nav school and then on to Wichita for B-47 ground school. None of us had any jet time so we flew T-33s for about six weeks and then started our B-47 flight training. My first B-47 flights were at Wichita.

BILL SLADE

Through 1949 and 1950 as George Birdsong, Lloyd Griffin, Bill Slade, and many of their contemporaries trained to become the first operational Stratojet crews world events were, as previously mentioned, taking an ominous turn. The advent of the Korean War in June 1950 no doubt played a role in dramatically increasing orders for Boeing's new jet bomber. By the end of 1952, 60 of the B-47Bs originally ordered in 1949 had been delivered, with the remainder soon to be completed. During this same period, however, several more contracts were issued, and in 1952 no less than 1,760 B/RB-47Bs were ordered. Ultimately, the order was paired down and most of these aircraft were actually completed as B-47Es. A total of 399 B model Stratojets were built.

The rapid succession of contracts issued and the resulting accumulation of orders which swelled Boeing's books was indicative of the urgency felt by SAC, the Air Force, and leaders in Washington D.C. to build a strategic bomber force equipped with the Stratojet . . . and to do it in a hurry. In fact, such was the demand for B-47s that the USAF decided that Boeing would not be able to supply Stratojets fast enough or in sufficient numbers. The result was the reinstitution of the well-known "Boeing-Vega-Douglas" pool which was formed during WWII to crank out thousands of critically needed B-17 Flying Fortresses. Lockheed and Douglas were once again pressed into service to build yet another urgently needed Boeing bomber.

Both companies would assemble and complete significant numbers of the aircraft at government-owned plants similar to Boeing-Wichita. Lockheed Stratojets would be put together at the government-owned plant in Marietta, Georgia, while Douglas-built versions would roll off the production line at the government's Tulsa, Oklahoma facility. Lockheed-built Stratojets were designated B-47-LM and Douglas models were B-47-DTs. Boeing's Wichita-fabricated aircraft took the designation, B-47-BW. Lockheed and Douglas received contracts in 1950 but did not begin production until 1953 with the arrival of com-

B-47Bs on the ramp at Boeing-Wichita undergoing final preparations for delivery. Using the same type of lighting employed by baseball stadiums, Boeing made round-the-clock shift work possible. It was just one of the efficiencies that allowed Boeing to crank out so many Stratojets so fast. (*National Archives*)

ponents supplied by Boeing-Wichita. Hence, just ten B-47Bs were built by Douglas and eight by Lockheed. Further production was immediately switched to the updated B-47E.

Upon completion of 10-25 school, graduates had their first chance to fly the B-47. They had a few hours in the T-33, but this would be their first experience in a large multiengined jet, one similar in size to the multiengined piston-powered bombers many of them had flown previously. Their initial impressions were overwhelmingly positive.

> Oh the acceleration it had on takeoff was fantastic. We had so much more thrust than we'd had in any propeller-driven aircraft. It really pushed you back in your seat. My first landing also made quite an impression on me. It was hard getting the correct attitude for landing. It was strange to me. You didn't really flare the airplane much. You just flew it on. And it was difficult to get slowed down. If you came in on final with an extra 10 knots, you needed another 1,000 feet of runway.
>
> BILL SLADE

> I entered the program so early that we didn't even fly the airplane in Wichita. We went right down to MacDill [AFB] where they checked us out in about two weeks. We replaced the original cadre of B-47A pilots and became IPs (instructor pilots) while they went to 10-25 school. We took our first few rides in the back seat just to get used to the airplane and get the feel of it. The next day we were in the front seat. There were a lot of systems and emergency procedures to learn, but once we did, it was fine. The acceleration on takeoff was the first thing that impressed me. The climb and the maneuverability were good and the ease of the controls was great with the hydraulic boost we had. It was tough to slow down, though. You had to fly the airspeeds exactly right on approach, I mean within a knot. If you were 2 knots over you'd go about 200 feet further down the runway. If you figured that your approach speed was say, 152 knots, you'd jostle your airspeed around til' it was just right. You'd put it right on 152 knots.
>
> GEORGE BIRDSONG

> Having the right approach speed was critical. At the time, we had to fly what was known as "best flare speed," and that was just a couple knots above stall speed. Soon, it was found that many people were landing short of the runways and rolling up to them. So they added about 6 knots to our approach speed and changed it to "best approach speed."
>
> BILL SLADE

While the original production order for 87 B-47Bs was being filled, General Electric was completing development of a more powerful J47 engine. J47-GE-23s rated at 5,800 pounds of thrust were fitted to the succeeding 302 B-47Bs and were retrofitted to the original 87 B models. The welcome increase in power helped slightly to offset the some of the added weight of the B-47B.

Even as B model Stratojets began to come off the production lines, efforts were under way to improve the airplane. Boeing and the Air Force identified several areas which could be refined. The refinements resulted in the B-47E, which made its first flight in January 1953. In the meantime, many of the B-47E upgrades trickled down to the B-47B. Modifications to the Bs began as early as 1952 and continued until 1957. By June 1953 production of the B-47B ceased and construction of the E model began. In the same month two programs, *High Noon* and *Ebb Tide,* were approved. These consecutive modification phases brought the B-47B up to B-47E standards. When the programs were complete in 1957, the B-47B in effect vanished, having been transformed to what were unofficially known as B-47B IIs. The B-47B was now indistinguishable from the B-47E.

As the 306th BW's three squadrons grew to full strength during 1952, the Wing began to learn just what it had in the Stratojet. Flight crews quickly realized that the B-47 was an aircraft which demanded attention. A number of dangers awaited the inattentive or unlucky crew. Early in the transition phase a small number of mishaps occurred and several aircraft were lost—some due to pilot error, and some due to freak accidents such as canopy failures. But the B-47 was fundamentally a safe aircraft if flown correctly. It was not the dangerous aircraft it has occasionally been portrayed as. A look at the mishap rate for the Stratojet confirms that it compared quite favorably with the accident rates of other early jets. An informal survey for this book of a wide array of B-47 pilots with experience in all of the variants of the aircraft bears out this finding. The Stratojet was well-built and safe.

> I can't say enough good things about it. I loved it! I didn't know anyone who flew it who didn't fall in love with it. It was a good, safe airplane. For example, early on we landed without the brake chute all the time until they [*SAC*] decided it was better for the brakes if we always deployed the chute. From then on, you had to use it. That was the rule. Remember, at the time, everyone was learning about the aircraft. It really was forgiving, though. You know that if a brand new pilot could fly one for any length of time without incident, it had to be forgiving. They made you be so careful with the aircraft. I knew that it could do things that SAC wouldn't allow you to do because I had done them! In the beginning at MacDill, we'd make our final approach turn at 500 feet and roll in at about 300 feet. It worked fine. But they decided pretty quickly that we had to stop that. We had to stay up at 1,000 feet and move our final back and make it more of a straight-in approach. It was a little safer but it was still safe to do it the other way.
>
> **LLOYD GRIFFIN**

A posed but evocative shot of the day-to-day routine at a Stratojet equipped SAC base. Fire crews were ever present in case trouble flared. This particular B-47B of the 321st BW at Pinecastle AFB, Florida looks fine though. (*National Archives*)

Safe though it was, the Stratojet did, like any aircraft, have limitations. One of the widely known hazards of the B-47 was a phenomenon known as *Coffin Corner.*

That was function of altitude, airspeed, and weight. Your airspeed and altitude curve looked just like a parabola. At the top, the curve would drop away within 3 or 4 knots on either side. A couple of knots too slow and you'd be in a stall. A couple of knots too fast would put you into a "buffet" or high-speed stall. I found Coffin Corner on a flight out of MacDill down to Puerto Rico. We were on a navigation training mission, navigating by shooting the stars. We ran into some weather and we just kept going up to get over it so we could keep the stars in view for the navigator. We kept climbing and we still weren't out of it. At 37 or 38,000 feet I ran into a little turbulence and that put me into one of those stalls. No matter what I did, I just couldn't get out of it. We just kept shuddering on down, losing altitude. I told the copilot and the navigator, "If we get down to 10,000 feet, we're leaving!" We pulled out of that stall at about 20,000 feet. If you were too high and too slow or too fast, you stalled. There was a very small margin for error. That was Coffin Corner.

BILL SLADE

In tandem with learning the flying characteristics of the B-47 306th crews were learning how to put the Stratojet to work. A major milestone in the process of making the aircraft operational was an exercise called *Sky Try*. Between January 22 and February 20, 1953 the 367th BS was tasked with pushing the Stratojet to its limit. The squadron flew their airplanes hard, evaluating everything from tactics and crew coordination to maintenance, logistics and security. The exercise was a check on all operational procedures in simulated combat conditions. Its importance was high, as the unit's first deployment was less than five months away.

It was a complete mission analysis. We were on the go around the clock, turning airplanes around as fast as we could, checking on supplies we'd need to keep the aircraft flying and testing crew proficiency and aircraft reliability. We also evaluated mission performance in terms of bombing and navigation. It was quick but concentrated training and a check on the aircraft to see what the requirements might be with regard to supplies and equipment.

LLOYD GRIFFIN

Sky Try was a success and preparations for the first deployment of the B-47 went full speed ahead. The impending deployment represented not only the first trip overseas for the Stratojet but the first overseas deployment of a full wing of jet bombers. The location chosen was England, RAF Fairford to be exact. Rotational deployments of B-29s and B-50s to various bases in the United Kingdom were now routine, having been under way since the late 1940s. But taking a wing of jet bombers to England required some special preparation. It was for just that reason that the 306th BW commanding officer, Colonel Michael McCoy, made a preliminary visit to England with his three squadron commanders, George Birdsong, Lloyd Griffin, and 368th BS C.O., Benny Close.

We flew in to Fairford and some other bases to set up a good jet instrument approach system for the B-47 for our wing and others that would follow. There was none at the time, of course. So we set up all the routine letdown procedures.

GEORGE BIRDSONG

Col. Mike McCoy and Capt. George Birdsong lead the 306th BW on the first B-47 deployment in June 1953 to RAF Fairford, England. (*National Archives*)

This B-47B is from the 3206th Test Wing at Eglin AFB, Fla. (*National Archives*)

A terrific study of a B-47 copilot at work. Just above, another B-47B flies forma-tion. This Stratojet appears to have no tail armament, suggesting perhaps that it could be a TB-47B. (*Boeing courtesy of Jack Wright*)

The quartet left for Fairford in early April. McCoy and Birdsong piloted one aircraft while Griffin and Close flew the other. They were about to make history. It would be the first over-seas visit by the Stratojet and, in the process, they would set a record.

We didn't plan on the record but we were the first ones to fly trans-Atlantic in a B-47 and it was so damn fast that every time you flew it somewhere—if you were the first one—you'd set a record.

GEORGE BIRDSONG

Both aircraft departed MacDill and flew to Limestone Air Force Base, Maine, pausing there overnight before beginning the trans-Atlantic leg to Fairford. This very same route would be used many times over by deploying Stratojet bomb wings in future years. The next morning the pair left Limestone and scorched along, reaching the United Kingdom in short order. A record was indeed set. But the story behind it has never been fully told. Inbound to Fairford McCoy and Birdsong ran into a slight difficulty.

A B-47B drops a load of 500-pound bombs from its short bomb bay. Though B-47 crews did drop live ordinance on occasion, this was an infrequent practice. Simulated bombing was much more common and RBS runs formed the bulk of their training. (*Boeing courtesy of Jack Wright*)

McCoy was wearing a survival suit made out of rubber that was giving him quite a bit of trouble in the airplane. I wasn't wearing one. Brize [*RAF Brize-Norton*] and Fairford are very close together, only about six miles apart. On our way in, I guess the navigator mistakenly identified Brize-Norton as Fairford. But I took a look at the map as we were on final approach and I said, "Boss, I think that's Brize-Norton, not Fairford!" He said, "Aw @#%*!!," and he landed and jettisoned the chute. I said, "See those B-50s on either side of the runway? That's the 43rd Bomb Group!" Before we could call the tower or anything, he took off again without permission and said, "This suit is killin' me! You [*Birdsong*] wanna make this landing?" So I landed it at Fairford. As soon as he got out of the airplane he took his knife and cut about half the suit off. He was red as a beat underneath! It had been itching and he was really trying to get on the ground in a hurry. He was an excellent pilot, though. We were first on the ground at Fairford, which is how they set the record, but Griffin and Close actually beat us there and they would've been first on the ground if they'd have landed when they got there. But they let McCoy land so he could get the record.

GEORGE BIRDSONG

Nevertheless, McCoy and Birdsong did set a record, completing the 3,120-mile transAtlantic leg in 5 hours, 38 minutes. The record would not stand for long, however. It would soon be eclipsed by other crews in the 306th. The Wing departed for the United Kingdom on June 3, 1953, with each squadron's 15 aircraft launching successively from MacDill. In turn, each stopped at Limestone AFB before continuing on to Fairford. Several crews broke

"Cheri-Lynn," one of the 306th BW B-47Bs, soars off the runway at MacDill AFB in June 1953, leaving for RAF Fairford, England, on the first-ever B-47 deployment. (*National Archives*)

McCoy's record on the trans-Atlantic leg, the last aircraft to land at Fairford setting a new mark at 5 hours, 21 minutes. The last trans-Atlantic speed record attained by a B-47 was set five months later in November, 1953 by a Stratojet going from Limestone to Brize-Norton with a time of 4 hours, 53 minutes.

KC-97s from the 306th's ARS (Aerial Refueling Squadron) accompanied the Stratojets to Fairford to support them during the deployment, and with all the players in place, the Wing began the first-ever B-47 deployment under the command of SAC's 7th Air Division. The 90-day TDY went quite well and the Stratojet's performance was good.

> The deployment was quite successful. We shook down our crews and accomplished all the training we wanted to. The Wing was on 24-hour alert and we flew training missions on the radar bombing sites and bombing ranges in the "Wash" of England and various missions to the continent to verify navigation procedures for certain locations and targets. General LeMay visited to see how we were doing. Everything went well and we came home confident and combat ready.
>
> **GEORGE BIRDSONG**

By the end of 1953 the B-47 was a fully operational, combat-ready bomber. Production was in high gear and Stratojets swelled the ranks of reequipping units. The 305th BW, the sister unit to the 306th at MacDill which had received its complement of B-47Bs began a three-month deployment to RAF Brize-Norton in September 1953 as the 306th returned home. The 22nd BW at March AFB became the third Wing to reequip with the B-47, and by the end of 1953 five more units had the new jet bomber. The B-47B had established the Stratojet as the most potent weapon in SAC's arsenal. The B-47E went further. It became, "The Backbone," of SAC.

Airpower on display. From the early 1950s until the early 1960s the Stratojet was SAC's main strategic bomber. (*National Archives*)

Stratojets on the flight line at Offut AFB, Nebraska in December 1955. Most of the bombers appear to be B-47Bs and Es. Curiously though, the aircraft in the foreground has no tail armament. (*National Archives*)

Floating over the runway threshold at Orlando AFB, Florida, this B-47B is just about to deploy its brake chute. (*National Archives*)

Specifications for the B-47B are as follows:

Performance

Maximum speed   608 mph/528 knots at
   16,300 ft 565 mph/491 knots at 35,000 ft
Cruise speed   498 mph/433 knots
Stall speed   177 mph/154 knots
Service ceiling   33,900 ft
Combat ceiling   40,800 ft
Initial climb rate   2,560 ft/min
Maximum climb rate   4,775 ft/min
Combat radius   1965 mi w/10,000-lb bombload
Maximum range   4,444 mi
Takeoff ground run   9,100 ft, 7,200 ft w/ATO

Dimensions

Wingspan   116 ft
Length   106 ft, 10 in
Height   27 ft, 11 in
Wing area   1,428 sq ft

Armament

Two .50 caliber machine guns

Bombload

18,000 lb

Weights

Empty   78,102 lb
Combat   122,650 lb
Gross   184,908 lb

Crew

Three: pilot, copilot/radio
   operator/gunner, bombardier/
   navigator

Serial Numbers

B-47B 49-2642 through 49-2646, 50-001 through 50-081, 51-2045, 51-2047 through 51-2356

KB-47G 50-040

YB-47F 50-069

YB-47C 50-082

XB-47D 51-2046, 51-2103

WB-47B 51-2115

DB-47B 51-2160, 51-2162, 51-2191, 51-2192, 51-2234

YDB-47B 51-2186

CL-52 51-2059

Lockheed B-47B-LMs 51-2197, 2204, 2210, 2217, 2224, 2231, 2237, 2243

Douglas B-47B-DTs 51-2141, 2150, 2155, 2160, 2165, 2170, 2175, 2180, 2185

B-47B Variants and Conversions

The advent of the B-47B in 1951 gave SAC its first truly capable jet bomber. However, the Air Force and Boeing were not resting on their laurels. Work was already under way to produce an improved version of the bomber. As B models began to fill the roles of SAC Bomb Wings, the Air Force realized it had something special. Units developing the aircraft's operational envelope found it well suited to its mission. At the same time, new technologies and missions were emerging. SAC found itself taking on these missions with a variety of rapidly aging aircraft, including B-29s and B-50s. Many of the new missions involved a certain amount of risk, and in a world where faster and newer jet fighters were proliferating, the performance disadvantage of these propeller-driven aircraft was potentially lethal. When the Air Force searched for alternative platforms to take on these missions, it quickly realized that its new jet bomber might fill the bill. The Stratojet was fast and large, with enough internal capacity to accommodate a variety of special-mission equipment or to serve as a flying test bed.

RB-47B

One of the special missions for which the B-47 was a candidate was the crucial business of aerial reconnaissance. By the early 1950s the mission had become supremely important. As tensions rose between the Soviet Union and the West, the gathering of accurate intelligence depended heavily on aerial reconnaissance. But with the development of a new generation of Russian jet fighters, photographic flights around the borders and occasionally into Soviet airspace had become very risky. The long-ranged RB-29's and RB-50's flying strategic reconnaissance missions were vulnerable. The jet-powered RB-45 was primarily employed for tactical reconnaissance, and while somewhat faster, was lacking in range and still didn't have the speed to keep it safe from intercepting jet fighters. Something faster, longer ranged, and more versatile was needed. The B-47 was speedy and had room for sensors and cameras.

Design work on a reconnaissance version began in 1951. It was decided that many of the upgrades due for the next bomber model of the Stratojet, the B-47E, should be incorporated in the RB-47B. A newer, more reliable A-5 FCS was installed, giving some piece of mind to the crew as they ventured into hostile areas. The still experimental, higher-thrust J47-GE-25 engine was to be used as well as a system of advanced photographic equipment.

It was projected, however, that the Reconnaissance version would not be ready until 1954. By that time the new B-47E would be operational, and since the RB-47 would share more in common with the E model than the B, the Air Force decided to designate the finished product the RB-47E. In the interim, SAC directed Boeing to modify 24 B-47Bs with an eight-camera sensor suite in a heated capsule mounted in the bomb bay. The aircraft would have only a daylight photographic coverage capability but would allow the command some reconnaissance capability earlier than otherwise possible. The RB-47B served with a number of units being employed exclusively to train crews between 1953 and 1955.

TB-47B

In early 1951 plans were made to establish a centralized B-47 flight training program. McConnel AFB, Kansas, was chosen as the training base and the 3520th Combat Crew Training Wing (CCTW) began operations there on June 5, 1951. Early graduates of 10-25

TB-47Bs from the 4347th CCTW fly in formation. (*National Archives*)

A TB-47B, its approach chute deployed, slows for landing. (*Boeing courtesy of Jack Wright*)

TOO MUCH RUDDER

An illustration typical of B-47 crews' humor from B-47 navigator Andy Labosky. (*Andrew Labosky*)

school had already been through transition training in the B-47B at Wichita. But when the 3520th stood up just across the ramp, the Air Force decided to adapt a number of B-47Bs specifically for crew transition training.

A total of 66 B-47Bs were transformed into TB-47Bs with two simple modifications. A fourth crew position was added for an instructor pilot or navigator-instructor as needed, and the aircraft's tail gun armament was removed. TB-47B modifications were executed at two facilities. Forty-eight B-47Bs were modified at Douglas' Tulsa plant and 18 more were modified by the Air Force itself at the Oklahoma City Air Materiel Area at Tinker, AFB.

The TB-47B did yeoman duty with the 3520th CCTW, providing transition training for the majority of B-47 crews until 1958 when SAC took responsibility for training Stratojet crews. The TB-47Bs were retained as part of the 4347th CCTW, also located at McConnel AFB.

MB-47B

The technologies which lead to the atomic bomb and those which lead to the Stratojet progressed almost simultaneously, and from the outset, the Stratojet had been designed as a nuclear bomber. With orders pouring in for B-47s it looked as if the Stratojet would be the main delivery system for nuclear weapons for some time to come. By 1950 when the first B-47A flew, the second generation of nuclear weapons was close at hand. Development of the hydrogen bomb was nearing completion, but employing this more powerful weapon presented a special problem. With a destructive force nearly 100 times as great as that of the atomic bomb, the hydrogen bomb was expected to produce such a huge explosion that any aircraft dropping it would be swallowed up in the blast. While nuclear weapons and the B-47 were maturing together, the technology necessary to field unmanned delivery systems such as guided missiles was still under development. Accordingly, the Air Force searched for an interim means of delivery.

Though the requisite technology for guided missiles was still a way off, the capability did exist to remotely pilot an unmanned aircraft to a target. The solution seemed feasible and SAC took a look at its inventory to find a suitable aircraft. The brand new B-47 fit the range and payload requirements and had sufficient space to incorporate the electronic equipment needed to guide the aircraft. Early in 1950 the Air Force decided to select one of the ten B-47As which Boeing would soon deliver for modification as a drone director aircraft with the designation DB-47A. Two examples of the upcoming B-47B would be modified to carry the hydrogen bomb as unmanned MB-47B drones.

The program formally began in April 1951 under the cryptic name *Brass Ring*. Many issues were unresolved, though. How would the MB-47s operate? Would they dive on their targets like a missile or drop the hydrogen bomb in the conventional manner and be vaporized in the explosion? Where would the MB-47s launch from and how would they reach their targets? The hydrogen bomb was expected to be large and heavy. Flying to targets a long way off would require inflight refueling of the drone, perhaps several times. It was suggested that the drones be manned by a crew who would carry out the refuelings and then bail out over allied territory after the last refueling. The MB-47B would then fly on to the target unmanned with guidance from autonomous stellar tracking systems or onboard automatic navigation. The challenge lay in developing an automatic guidance and bombing system, one which would also be impervious to jamming. It soon became clear just how difficult this would be, and the Air Force began to ponder the practicality of having the director DB-47 accompany the MB-47B all the way to its target.

The program came to nothing, however. In the meantime a much simpler solution had arisen. A parachute-equipped hydrogen bomb could be dropped normally from a bomber, leaving time enough for the airplane to clear the blast area. Moreover, this could be made to work with either of SAC's existing heavy bombers, the B-36 or B-47. Consequently, no MB-47Bs were ever built and the Brass Ring program was canceled in April 1953.

DB-47B

In the late 1940s a number of weapons systems were evolving simultaneously. The development of air-launched guided missiles was not far behind the introduction of the B-47, and the Stratojet was immediately considered as one of several platforms for launching these weapons. In 1949 development of the Bell GAM-63 Rascal air-to-surface missile was initiated.

Powered by a liquid fuel rocket producing 4,000 pounds of thrust, the 13,000-pound missile could carry a 3,000-pound nuclear warhead up to 100 miles at Mach 2.95. Once launched, the Rascal was directed by a guidance system aboard the DB-47B. Bell's Radar Scanning Link guidance system for the GAM-63 resulted in the acronym *Rascal*. Four bombers were originally considered as launch platforms for the Rascal, including the B-36, B-47, B-52, and B-60. By 1952, only the B-47 and B-52 were in the running, and with the B-52 still in experimental development the B-47 was given first priority. The GAM-63 was attached by special fittings to the starboard side of the DB-47B's fuselage.

The Rascal was designed to be a "stand off" weapon, allowing B-47 crews to launch it from as far away as 100 miles, thus minimizing the crew's exposure to antiaircraft defenses in the target area. The launch sequence would begin with the release of the missile from its mounts. Dropping below the aircraft, the Rascal's rocket motor would fire once it was a safe distance away. A modified B-50D made the first successful launch of the Rascal on September 30, 1952. Over the course of 1953 a B-47B was modified to carry the GAM-63 with the designation YDB-47B. Two early B-47Es were also modified in anticipation of adapting many of the newer E models to this mission.

Several problems plagued the program. First, though the Air Force was solidly in favor of the project, SAC wasn't and registered its doubts loudly and officially. The command's reservations included concerns about degradation in aircraft performance resulting from the installation of the missile and a suspicion that the guidance system would be troublesome and inaccurate. Furthermore, the idea of adding the necessary electronic equipment to an already complex combination of aircraft systems aboard the B-47 was not appealing to SAC. Just as worrying was the expense of the million-dollar-per-aircraft modifications and the investment in crew training that would be required.

However, by 1955, the Air Force was still behind the project, and despite SAC's objections, decided to move forward. Accordingly, 30 B-47Bs were set aside for modification with a further 44 to follow. The first successful Rascal launch from a DB-47 was made in July 1955 from one of the two DB-47Es. Over the next two years the program slowly ran out of steam. Orders for additional DB-47Es were canceled and it was decided that only one DB-47/GAM-63 squadron (the 445th BS of the 321st BW at Pinecastle, AFB) would be formed. Despite the delivery of the first production Rascal to the 445th in October of 1957, the program was canceled within a year. Reports of the missile's effectiveness are contradictory, but by 1958 it was nearly obsolete anyway, having been technologically surpassed by the *Hound Dog* and *Quail* missile systems, which soon became operational on the B-52. Ultimately, 74 DB-47Bs were completed.

WB-47B

One of the few B-47s in the Air Force inventory not operated by SAC was a single B-47B modified in 1956 as a weather reconnaissance aircraft for Military Air Transport Service's

Air Weather Service. After the devastation of hurricanes Carol and Hazel in 1954, Congress requested that a more modern and capable hurricane hunter be built to replace the aging WB-50s in service. The result was the WB-47B. The aircraft was to be used to penetrate hurricanes and fly a diverse range of weather-related missions. But its effectiveness, especially as a hurricane hunter, was limited. The WB-47B and later WB-47Es were the only jets ever to fly the hurricane mission but were hampered by their high cruise speed. Just as you slow down to drive over a speed bump, aircraft are flown as slowly as possible in the severe turbulence associated with a hurricane. For this reason, the WB-47B/E could not penetrate the interior of a hurricane and was restricted to skirting the edges of the storm.

In 1958 the sole WB-47B was tapped to do research and data collection with the first ever weather satellite, the Tiros II. The aircraft was assigned to the 55th WRS of the 9th Weather Reconnaissance Group, serving from November 1957 to November 1965 when it was replaced by the newer WB-47E.

YB-47C

Throughout their service life most military aircraft go through a never ending process of refinement. So it was with the Stratojet. Efforts aimed at improving its performance were frequent, especially in its early years of service. One of the most interesting proposals entailed a major revision of the B-47's powerplants. Weight gains in the form of added fuel capacity and mission equipment left the B model Stratojet slightly underpowered. The newer J47-GE-23 engines helped but didn't completely close the gap. Boeing and the Air Force wanted more power for the B-47, and in 1950 proposed a version featuring four, instead of six, jet engines. Initially, they chose Allison's 10,090-pound thrust J71-A-5 turbojets to power the new Stratojet. The conversion was at first thought different enough to

An artist's concept of the never-produced B-47C. (*National Archives*)

warrant a completely new designation. The aircraft would be known as the YB-56. A reconnaissance version, the RB-56, was also planned. The 88th B-47B was chosen for conversion as a prototype.

The Air Force had high hopes for the YB-56, believing it could be the ultimate Stratojet. However, problems with the J71 engines cropped up and they were dropped from the project. In the meantime, the prototype was redesignated the YB-47C, the earlier nomenclature having been abandoned in acknowledgment of the overwhelming commonalties the new aircraft would have with its predecessors. In search of a suitable engine, the Air Force evaluated the Pratt & Whitney J57. The J57 was still in development though and was, in any case, slated for use on the upcoming B-52. Despite the promise of greatly improved performance the decision was made to terminate the project in December 1952. The B-47B set aside for the program was never converted.

XB-47D

Another of the roles which the Air Force quickly realized the B-47 might be suited to was as a flying test bed for the development of new technologies. The XB-47D was just that.

In the late 1940s and early 1950s the USAF was interested in investigating the viability of turboprop-powered aircraft. One application of particular interest was the possibility of developing a high-speed, long-range turboprop bomber. In the first of several different engine test programs that the Stratojet would be involved in throughout its career, the Air Force requested that two B-47Bs be converted to flying test beds to examine the feasibility

One of the most interesting modifications ever made to the Stratojet was its transformation from pure jet to turboprop/jet configuration. Two B-47Bs were converted to XB-47Ds to study the feasibility of a turboprop bomber. Surprisingly, the turboprop/jet combination yielded similar performance to the conventional version. Readily apparent in this shot are the huge paddle blades that provided thrust for the Curtiss-Wright T49 engines. (*National Archives*)

of turboprop engines. Evaluations would be made of the turboprop/swept wing combination and the turboprop/turbojet blend.

The conversion program began in April 1951 with Boeing removing the inboard J47-GE-23s of the B-47B and replacing them with single Curtiss-Wright YT49-W-1 turboprops of 9,710 shaft horsepower each. Massive four-bladed propellers, 15 feet in diameter, featuring 24-inch-wide paddle blades converted the T49's power to thrust. A consequence of the turboprop installation was modification of the wing flaps to allow proper clearance of the engine nacelles. Changes in the cockpit included different instrumentation to manage the two turboprops and two J47s.

It was not until August 26, 1955 that the first XB-47D made its maiden flight. Problems with the T49 and with the engine/propeller combination and equipment shortages delayed the program. Once in the air, the XB-47D displayed similar performance to the conventional B-47. Landing performance was actually enhanced with the addition of the engine's fully reversible propellers to the normal brake chute. With a top speed of 597 mph at 13,500 feet it is claimed that the XB-47D achieved the fastest speed ever for a propeller-driven aircraft in level flight. No range figures are available for the aircraft.

After a significant number of successful test flights were made, the Air Force decided to cancel the project. No further examples were built.

Specifications for the XB-47D are as follows:

Performance

Maximum speed   597 mph
     at 13,500 ft
Service ceiling   33,750 ft.
Initial climb rate   2,910 ft/min

Dimensions

Wingspan   116 ft
Length   108 ft
Height   28 ft
Wing area   1,428 sq ft

Armament

None

Powerplants

Two Wright YT49-W-1 turboprops of 9,710 shaft hp each
Two J47-GE-23 turbojets of 5,800 lb static thrust each

Weights

Empty   79,800 lb
Gross takeoff   184,428 lb

Crew

Three: pilot, copilot/radio operator, gunner/
     navigator/bombardier

KB-47G / YB-47F

Among the most interesting adaptations of the Stratojet was its conversion as an aerial tanker. The effort was launched in 1953. Its goals were twofold. In the early 1950s SAC was still evaluating the two extant methods of aerial refueling, Boeing's *Flying Boom system* and the *British Hose and Drogue method*. The Boom system, which Boeing introduced with its KB-29P, worked well but SAC wanted to test alternatives, knowing that it would soon have a large force of thirsty jet bombers. In addition, the command wanted an aerial refueler that was speedier than the existing KB-29s, KB-50s, and the newly operational KC-97. Not only were SAC's bombers thirsty, they were fast. The speed mismatch between them and the propeller-driven aerial tankers was significant. Refueling a B-47 behind a KC-97 could be a tricky business.

The XB-47D being readied for taxi tests. Both cockpit instruments and flaps had to be modified to accommodate the Curtiss-Wright T49s. (*National Archives*)

> We were riding the edge of a stall the whole time we were behind a KC-97. The airplane was shaking. If you were really gonna take on a big load of gas, the tanker would have to start downhill and head for the ground. We had to put them in a descent because as we got heavier there was no way we could maintain position without stalling.
>
> **JOHN IRVING, PILOT, 40TH BW, 1954–1959**

To evaluate the B-47 as a tanker a single B-47B was converted to carry a British supplied hose and reel in its bomb bay. The tanker was given the designation KB-47G. A full evaluation of the Hose and Drogue method for the Stratojet was ready to get under way with the conversion of another B-47B to receive fuel. A probe was installed in the nose YB-47F just ahead of the cockpit. The test program was undertaken at the Wright Air Development Center at Patterson Field by Air Research and Development Command.

> I was in the Directorate of Flight and All Weather Testing, assigned to Bomber Test when the program got under way. I flew both "Ma" [*KB-47G*] and "Paw" [*YB-47F*]. Refueling by that method wasn't too difficult, but it wasn't really that practical, particularly with the B-47. It gobbled fuel itself and couldn't carry enough extra to give another airplane a decent off-load. The only extra tankage you could put on a B-47 was the drop tanks, and that really wasn't a heck of a lot of fuel. There were no extra internal fuel tanks because there was no place to put them.
>
> The refueling reel was mounted in the bomb bay. The hose we had was the biggest hose they used. The copilot operated it from his cockpit. He opened a separate set of doors—not the bomb bay doors—just ahead of the rear gear and automatically unreeled the hose.
>
> I made the last flight with the tanker. A 101 [*F-101 Voodoo*] was workin' the hose. Normally, if the receiver aircraft overran the hose, the reel would retract and take up the slack.

Then, if the aircraft backed off, the reel would feed the hose back out. What happened was the F-101 overran the hose, and when he backed off the hose reeled out to its full extension. What nobody knew was that the hose came off the reel. When it wouldn't retract we just assumed that the reel wouldn't take it up. I had to decide what to do. I didn't want to land because bringing the airplane in dragging the hose would've been a big problem. I was afraid that if we landed that way the rear gear would step on that mother and tear the whole rig out of the airplane. We were lucky to get rid of the hose. Three rollers guided it out through the doors. When the hose came off of the reel the brass collar on the end of it jammed in the rollers. Amazingly, there was a hammer, a screwdriver, and a pair of pliers aboard the aircraft. I've no idea why. The copilot got out of his seat, went down into the bomb bay, and finally drove the axle pin out of one of the rollers. That released the hose. While he was doing that I was circling Lake Erie. The hose came free and fell into the lake and I'm sure its still down there. That was our last flight.

CHARLES ANDERSON, TEST PILOT, KB-47G/YB-47F, 1953–1954

Refueling tests were carried out with the F-84G and F-101 in addition to the YB-47F. But making the Stratojet a tanker did not prove practical. It could not carry enough gas to refuel jet bombers. The program was discontinued in 1954. However, a number of Air Force aircraft, especially fighters, continued with the probe and drogue system and the program was briefly revived in 1956. Tactical Air Command's new Century series of fighters encountered difficulties in refueling from the command's KB-50s and the Air Force evaluated a proposal to develop a two-drogue prototype B-47 tanker. Conversion costs proved too high, though, and the program was officially canceled in July 1957.

The one and only experimental KB-47G refuels the single experimental YB-47F via the "Hose and Drogue" method. Note the rear gear of the YB-47F is extended to slow its approach to the KB-47G. (*Mark Natola*)

Looking for a way to refuel high-speed aircraft such as fighters and the B-47 itself, the Air Force converted two B-47Bs to KB-47G tanker and YB-47F receiver aircraft using the British "Hose and Drogue" method. "Ma" (the KB-47G) and "Pa" (the YB-47F) proved to be impractical, however, because the B-47 couldn't carry enough fuel to offload the quantities required by large aircraft like the Stratojet itself. (*Mark Natola*)

CL-52

With the advent of the Convair B-36 a policy was begun of building and operating strategic bombers for national service only. No longer would strategic bombers be exported. B-17s, B-24s, B-29s, and B-50s all saw service in the air forces of foreign nations, but beginning with the B-36, no future strategic bomber would ever be in service abroad. There was one exception, however. A single Stratojet did briefly serve outside the United States as a flying test bed.

In the mid-1950s the British aircraft manufacturer, Avro, was developing a new fighter for the Royal Canadian Air Force (RCAF), the CF-105 Arrow, a long-range interceptor. Twenty-thousand-pound static thrust Orenda Iroquois turbojet engines were chosen to power the CF-105. In flight testing of the engines had yet to be accomplished, and a search was begun to find a suitable flying test bed. Finding nothing elsewhere that would suit their needs, Avro and the CAF looked to the United States for an appropriate aircraft. Subsequently, they struck a deal with the USAF for the loan of a B-47. Early in 1956 B-47B no. 51-2059 was transferred to the RCAF who bailed the aircraft to Canadair for modification. The company designation for the aircraft was CL-52.

I went through the aircraft commander course on the B-47 in 1956 at McConnel, AFB. My copilot/observer went as well. Avro had made arrangements to get an airplane from the USAF. They bailed it to the RCAF who then lent it to Avro/Canadair to do the testing with.

It was put in RCAF markings and red dayglo paint with a red ring around the fuselage and red tail surfaces. They chose the B-47 as a test bed because they'd looked at everything around and it was the only aircraft available that was big enough to absorb the thrust of the Orenda engine.

MIKE COOPER-SLIPPER, AVRO TEST PILOT, 1948–1960/CHIEF TEST PILOT FOR CL-52/

ORENDA IROQUOIS ENGINE PROGRAM, 1956–1959

After looking at several alternative locations Canadair and Avro chose to locate the Iroquois on the starboard side of the fuselage below the horizontal stabilizer. The engine pod was 30 feet long and 6 feet in diameter.

It was put on all wrong. It was on the starboard side and the tailpipe was towed out instead of in, and that gave the aircraft quite a "turning moment." You couldn't hold it straight with the rudder. You had to have everything throttled back, except engines number 1 and 2. Number 1 was at cruise power and number 2 was slightly throttled back. Numbers 3, 4, 5, and 6 were throttled all the way back.

They located the engine there because there was nowhere else to put it. There wasn't enough ground clearance to put it in the bomb bay. They could've put a retracting mount in the bomb bay but it would've been very complex and added a lot of weight. You couldn't put

Avro test pilot Mike Cooper-Slipper takes the only B-47 ever to serve outside the United States, the CL-52, aloft. The CL-52 was a test bed for the Orenda Iroquois Turbojet, proposed for Avro's CF-100 fighter/bomber. The Iroquois is mounted on the left wing external fuel tank pylon. Reportedly, use of the engine on takeoff made the CL-52 almost impossible to control. (*Mike-Cooper Slipper*)

it on the wing. So they decided the best place for it was under the tail. It wasn't a very good idea, really. It worked once you got up on power and running along but it was hard during acceleration. You couldn't keep up with the speed of acceleration of the Orenda engine. We had 32,000 pounds of thrust available with the afterburner. We did about 35 hours of testing. Up to about 25,000 feet it was all right. Above 25,000 feet it was too much. You just couldn't control the airplane with the engine operating. You had to pay a lot of attention to it, and you had to fine tune power settings to get the airplane flying straight. When you went above 25,000 feet a normal B-47 had a yaw damper that took care of the airplane's "Dutch Roll" problem. But with the engine at power the yaw damper couldn't cope. The CL-52 served half its purpose very well, getting the engine running and setting it at certain power settings at different altitudes. But you couldn't do accelerations at high altitude. The aircraft would just start to roll. Nevertheless, it [*B-47*] was the best airplane in the world for the job at the time.

MIKE COOPER-SLIPPER

Unfortunately, all the testing was for naught. After a host of delays and problems, both financial and political, the Avro CF-105 was canceled by the Canadian government on February 20, 1959. The CL-52/B-47B was returned to the United States and flown to Davis-Monthan AFB to be scrapped.

B-47E

If there was one classic version of the Stratojet, it was the B-47E. The E model was the "backbone of SAC," bringing the stature and importance of the command to a new level. Between 1953 and 1957 no less than 1,341 B-47Es joined SAC's inventory. The burgeoning population of B-47E's expanded to fill a score of air force bases in the continental United States and around the globe. A literal army of thousands of men flew, supported, and supervised the operations of E model Stratojets. The B-47E and SAC were synonymous. From the early 1950s to the early 1960s this aircraft was at the forefront of America's strategic deterrent forces. It was the major production version of the Stratojet, so much improved over its predecessors that the bulk of the B-47B fleet was brought up to E model standards. At any given time of day, on any day throughout those years, B-47Es were standing alert somewhere. Initially the force was on 24-hour alert, then a third of the force was on 15-minute alert, then half of the total force was on 15-minute alert, "cocked" and ready for action. The B-47E won four of SAC's highly competitive Bombing and Navigation competitions, set speed and endurance records, and was even featured along with B-47Bs in a Hollywood movie. By the end of 1956 SAC could claim 27 fully operational B-47 Wings with more than 1,200 combat-ready crews and over 1,300 Stratojets, nearly all of which were B-47Es. The Stratojet program represents the largest production of heavy bombers since the end of World War II and the B-47E was the definitive Stratojet.

Improvements to the new model were many. By the time the B-47E first flew on January 30, 1953 Boeing engineers and the Air Force had accumulated more than 5 years worth of flight test and operational data. Though the Stratojet program was still relatively young, the experience gained with the XB-47, B-47A, and B-47B pointed to a number of areas which could be refined. Modifications were on-going throughout the 4-year B-47E production run and occurred in four phases. The upgrades added in each phase were retrofitted to preceding aircraft.

Phase I Modifications

Phase I modifications began with the installation of more powerful J47-GE-25 engines. The basic dash 25 engine produced 5,970 pounds of static or dry thrust. However, General Elec-

B-47E-LMs. These freshly completed Lockheed built Stratojets are parked on the ramp at the government-owned Lockheed plant in Marietta, Georgia, awaiting dispersal to active units. (*National Archives*)

The lines of the B-47E were classic Stratojet, uncluttered and free of the bumps and bulges found on later reconnaissance versions of the aircraft. (*National Archives*)

tric quickly followed the dash 25 with the dash 25A, which, with the aid of water/alcohol injection, could produce 7,200 pounds of thrust for takeoff. This was a welcome addition indeed and helped to launch the Stratojet more quickly and safely. Early versions of the B-47E tipped the scales at just over 200,000 pounds at maximum takeoff weight. Unfortunately, later production E models gained even more weight due to a number of modifications added in succeeding phases. The extra weight once again swallowed up most of the increase in power. Still, the addition of another type of thrust augmentation in the form of water/alcohol injection did provide flight crews with some measure of comfort, especially when launching at heavy takeoff weights at high-density altitudes on hot summer days.

This type of thrust augmentation worked by spraying a water/alcohol mixture into the engine's combustion chambers and tailpipes, thereby increasing the mass flow of gases from the exhaust nozzles and, accordingly, the amount of thrust produced. Fifty-two hundred and ninety-six pounds of water were pumped in a period of 80 seconds. The pressure of the water/alcohol was simultaneously used to raise a small tab at the discharge end of the tailpipe. This decreased the cross-sectional area of the nozzle, raising discharge velocity and mass. The water/alcohol also provided cooling so that exhaust gas temperatures were not exceeded during this operation.

Having lined up on the runway for takeoff, pilots would advance their throttles to 100 percent power. After arming the system, the water/alcohol injection was turned on by flipping a switch on the right-hand side of the cockpit. This was done coincident with brake release. In the nose, the navigator began timing acceleration. After a precalculated number of seconds the navigator would announce the end of acceleration timing, at which point the pilot would check his indicated airspeed to ensure that the aircraft had acquired sufficient velocity for takeoff. Once decision speed had been reached successfully, the pilot would raise the nose and begin to climb at a fairly flat angle. Rising away from the runway, the boost ran out and pilots would actuate a "stop and drain" switch to make sure that all of the solution was gone before ascending into the low temperatures at higher altitude.

A B-47E poses, showing its fine lines. (*National Archives*)

The 306th BW under the command of Col. Michael McCoy and squadron commanders George Bird-song, Lloyd Griffin, and Benny Close, played a pioneering role in B-47 operations. Here, early model B-47Es of the 306th are being readied for flight on the flight line at MacDill AFB. (*National Archives*)

The inboard twin J-47-GE-25s on an early B-47E produced 6000 pounds of static thrust. Note the clever integration of the landing light along with the engine nacelles. (*National Archives*)

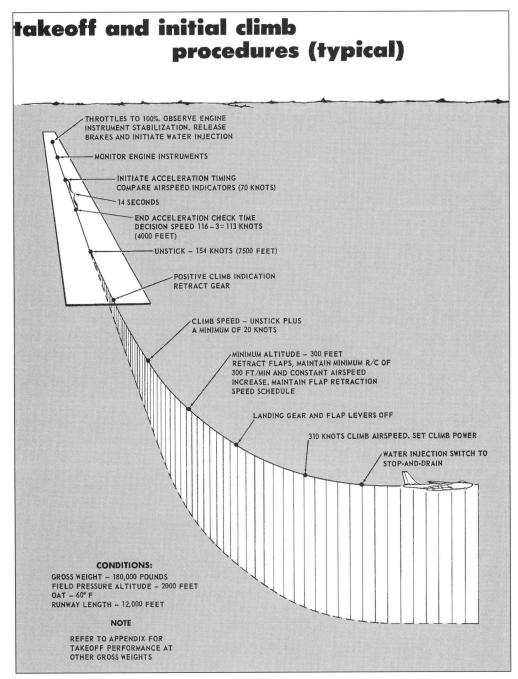

takeoff and initial climb procedures (typical)

THROTTLES TO 100%. OBSERVE ENGINE INSTRUMENT STABILIZATION. RELEASE BRAKES AND INITIATE WATER INJECTION

MONITOR ENGINE INSTRUMENTS

INITIATE ACCELERATION TIMING
COMPARE AIRSPEED INDICATORS (70 KNOTS)

14 SECONDS

END ACCELERATION CHECK TIME
DECISION SPEED 116 – 3 = 113 KNOTS
(4000 FEET)

UNSTICK – 154 KNOTS (7500 FEET)

POSITIVE CLIMB INDICATION
RETRACT GEAR

CLIMB SPEED – UNSTICK PLUS A MINIMUM OF 20 KNOTS

MINIMUM ALTITUDE – 300 FEET
RETRACT FLAPS, MAINTAIN MINIMUM R/C OF
300 FT/MIN AND CONSTANT AIRSPEED
INCREASE. MAINTAIN FLAP RETRACTION
SPEED SCHEDULE

LANDING GEAR AND FLAP LEVERS OFF

310 KNOTS CLIMB AIRSPEED. SET CLIMB POWER

WATER INJECTION SWITCH TO
STOP-AND-DRAIN

CONDITIONS:
GROSS WEIGHT – 180,000 POUNDS
FIELD PRESSURE ALTITUDE – 2000 FEET
OAT – 60° F
RUNWAY LENGTH – 12,000 FEET

NOTE

REFER TO APPENDIX FOR
TAKEOFF PERFORMANCE AT
OTHER GROSS WEIGHTS

B-47E takeoff and initial climb procedures. Notice the inclusion of water/alcohol injection procedures. Using this method of augmenting thrust was very common. (*Jack Wright*)

I flew B-47Es with the 509th BW at Walker AFB, New Mexico for a couple of years in the late 1950s. We made mostly water-augmented takeoffs because, as I recall, it increased takeoff thrust about 8 percent and we needed any edge we could get. We had a 13,500-foot runway with a field elevation of 3,777 feet and we loaded the aircraft for a 10,000-foot takeoff roll. The water lasted until we got up to about 300 feet, I think about 2 minutes. The B-47 had a very flat climb out as we accelerated to climb speed which was approximately 275 knots, from about 165 knots at takeoff. The water left heavy black smoke on the runway. If we made formation takeoffs with our normal one-minute separation you would not be able to see the aircraft ahead of you on the runway after a couple of departures. So you had to wait until the aircraft in front of you called airborne or until you could visually observe him climbing. SAC had an accident, as I recall, where one aircraft aborted takeoff and the following aircraft ran into him on the takeoff roll.

TOM STANTON, PILOT, 509TH BW 1956-1958

Indeed, the use of water augmentation during Minimum Interval Takeoffs (MITO) was a cause of concern among flight crews. MITO was developed for SAC's Stratojets in response to a growing threat from Soviet ICBMs in the mid-1950s. Targeted on U.S. air bases worldwide, the ICBMs had the effect of substantially reducing the reaction time SAC's alert force had to launch aircraft. To get B-47s in the air more quickly, the command began experimenting with shorter and shorter intervals between takeoffs under project *Open Road.* Doubts were great at first, and many thought that the turbulence and smoke left by departing Stratojets would allow for an interval of no less than 1 minute. Eventually, the interval was whittled down considerably, and by 1960 SAC B-47s were roaring off the runway every 15 seconds safely. However, bombers making MITO takeoffs using water augmentation or ATO or both did present dangers. Many pilots remarked that they couldn't even see the runway during MITO takeoffs with either form of thrust augmentation. To aid flight crews during the takeoff roll, SAC painted white guidance lines down the length of many of its runways. But as the crews were quick to point out, even if you could stay lined up on the runway, the little white lines wouldn't help you see another aircraft on the runway. If an aircraft aborted takeoff in front of you, you might never see it.

Dangers awaited crews making normal takeoffs as well. Losing an engine on takeoff is bad news in any multiengine airplane, but it was even more serious in the power-challenged B-47E.

A B-47E in flight. (*National Archives*)

A B-47B roars overhead making an ATO takeoff. Its 18-bottle internal ATO can be clearly seen. The nine units on either side of the fuselage swung outward to keep the flame they produced away from the skin of the aircraft. (*National Archives*)

Blast off! A B-47B makes an ATO takeoff using its 18-bottle internal system. Ever-increasing gains in weight meant that thrust augmentation was often needed to shorten takeoff runs and get Stratojets into the air quickly. (*National Archives*)

An early B-47E makes a maximum performance takeoff using both types of thrust augmentation, ATO, and water/alcohol injection. This E model employs the newer 30-unit external, jettisonable "horse collar" ATO rack. The thick black smoke belching from its engines is an indicator that water/alcohol injection is also being used. Stratojets on 15-minute alerts often had the option of using one type of augmentation or the other, or both together. (*National Archives*)

40th BW aircraft commander Capt. Robert Hanaway sails along in the pilot's seat of his B-47E. (*Robert Hanaway*)

A wonderful view back to the copilot's position in Robert Hanaway's 40th BW B-47E. Note the face reflected in the canopy.

B-47Es of the 40th BW, 44th BS on the Smoky Hill AFB flight line. Crew chief Don Carey and Lou Stevens work on aircraft 52-546. (*Don Carey*)

B-47Es of the 44th BS gleam in the sun at high altitude. The B-47 flew well in formation. (*Robert Hanaway*)

44th BS B-47Es from Smoky Hill AFB, Kansas in a gentle left bank (*Robert Hanaway*)

An RB-47H being readied for flight at Forbes AFB presents an interesting head-on view. (*National Archives*)

A "Mod-44" or "Silverking" RB-47H on the flight line at Forbes AFB. These updated Hs featured many new pieces of equipment including the pylon-mounted ALD-4 receiver pod. Here, the relative size of the ALD-4 is apparent when compared with the size of the ground crewman standing nearby. (*Bruce Bailey*)

The "Silverking" RB-47H's of the 55th SRW flew with both the 38th and 343rd SRSs. (*William Bateman*)

An ERB-47H of the 343rd SRS, 55th SRW. The ERB-47H was a more specialized version of the RB-47H. The two aircraft often flew together hunting for electronic data. The ERB-47H had a crew of five with only two EWOs as opposed to the six-man crew in the RB-47H. (*Jim Nelson*)

This EB-47E(TT) is one of only three that were completed. Its smaller "chopstick" antennas are no different than the larger "towel rack" type seen in other photos. It was found that the antennas did not need the protective white covers that fit over them. Consequently, crews removed them and benefited from lower drag. (*Bruce Bailey*)

The single WB-47B "Hurricane Hunter" which flew with the 9th Weather Reconnaissance Wing was perhaps the most colorful Stratojet ever in service. (*Courtesy of John Tegler*)

Two of the three B-47s ever to serve with the U.S. Navy. These two EB-47Es were flown by Douglas Aircraft Co. flight crews in support of the Navy's Fleet Electronic Warfare Support Group. They provided ECM support during fleet exercises. Here, the former 376th BW B-47Es sit on the ramp at NAS Point Mugu where they were based. (*Mark Natola*)

The last RB-47H to be retired from SAC was soon brought out of retirement to serve as an avionics test bed for General Dynamics F/B-111. This most distinctive H model was instantly recognizable by its F-111-like needle nose. (*Mark Natola*)

The last flight of a B-47: "Spirit" makes its way to the Castle Air Museum escorted by a T-33 chase plane.

A good study of a B-47E from below. Both gear have retracted after takeoff and the gear doors are just closing. (*Boeing, courtesy of Jack Wright*)

An illustration from navigator Andy Labosky which, though joking, illustrates well the apprehension Minimum Interval Takeoffs could produce. These takeoffs were hazardous because of the close spacing they required. B-47 pilots literally had to feel their way along the runway because the smoke from aircraft departing seconds before them left visibility virtually at zero. (*Andrew Labosky*)

A takeoff in the B-47E at, say, 216,000 pounds was long, long by even the standards of the day in the early 1960s. The 10,000-foot-plus ride to get to the 165-knot unstick was a perilous journey. An unusual phenomenon was for one of the outboard engines to fail due to what was thought to be a temporary fuel interruption. Boeing came out with a little news booklet to be distributed called *Engine Out Pedals*. The pamphlet stated that at unstick you had just 1.7 seconds to apply full opposite rudder in the event of an engine flame out or roll due to yaw would rear its ugly head. The upshot of that was that three crew members went up in a tremendous fireball.

DAVE RUSSELL, PILOT, 335TH SRS, 55TH SRW & 45TH BS, 40TH BW 1961–1963

Landing the Stratojet presented another set of challenges. Pilots could choose from several different letdown procedures, including gear-down and gear-up descents. Each of these were carefully executed to bring the aircraft down to approach altitude at just the right airspeed. Once on final, a careful ballet of power and airspeed management was played out, it being necessary to maintain sufficient power for a go-around while simultaneously slowing the slippery bomber enough for a safe touchdown. Measures were taken to aid the B-47E in this delicate dance. To help slow the aircraft, a 16-foot approach chute which could be deployed on final to increase drag while allowing higher engine RPM to be maintained was incorporated. Adding to the margin of safety was an antiskid braking system which was especially useful on rain-slicked runways.

Actually making contact with the runway demanded as much concentration as the preceding phases of the approach. The aircraft's tandem gear dictated that the Stratojet return to earth at the same attitude in which it taxied and took off. Getting it right was tricky. If the pilot allowed the front gear to make contact with the runway before the rear gear, the aircraft would begin to "porpoise." The only way to recover was to add power and take it around for another approach. If the pilot attempted to save the landing after the porpois-

A 376th BW B-47E on the tarmac at Barksdale AFB, LA. This early E model still has an internal ATO system. (*National Archives*)

Its brake chute billowing, a B-47E just after touchdown rolls out on the runway at Philadelphia International Airport, arriving for the 1953 National Air Show. (*National Archives*)

This B-47E lands with both its drag and brake chute trailing at Philadelphia International Airport for the National Airshow in 1955. (*National Archives*)

An early B-47E makes an ATO takeoff using a 30-bottle external, "horse collar" rack. (*Boeing, courtesy of Jack Wright*)

ing started, the springback only became more violent with the result that the aircraft would bounce higher in the air, stall, and fall off on one wing, then cartwheel.

Next in the series of Phase I modifications was the switch from internal 18 bottle ATO units to the jettisonable external 33 bottle "Split V" or "Horse Collar" racks previously described. A few of the earliest production E models retained internal ATO units, but the switch to external racks was made quickly and the overwhelming majority of B-47Es no longer featured ATO ports on either side of the fuselage above the rear gear, having instead a clean smooth skin. In fact, this is one way to distinguish the B-47E from its predecessors. The alert force of B-47s operating under SAC's Emergency War Order often had both forms of thrust augmentation, water/alcohol injection and ATO, available. Frequently, the two were used in concert with the water/alcohol being kicked in at the beginning of the takeoff roll followed by the activation of ATO several seconds later to propel the alert bombers skyward quickly.

Phase I B-47Es got a more potent "stinger," the much improved A-5/MD-4 FCS which featured a gun-laying radar and two M24A1 20-millimeter cannons, firing 350 rounds a piece. The system could track targets at a maximum range of 1,500 yards with improved accuracy and deliver a better defensive punch through the 20-millimeter cannons. Later, an upgraded AN/APS-32 gun laying radar was added.

Finally, the B-47E-I brought with it the welcome return of ejection seats for each of the three crewmen. Both the pilot and copilot ejected upward over the tail in the conventional manner. But the navigator ejected downward through an escape hatch below his position in the nose. The Stratojet was certainly one of, if not the first, aircraft to employ this method

A 40th BW ground crew on deployment to Lakenheath, England including Sgt. Ken Stansel, Airman Frank Bratjan, and Airman Paul Frye, explain the rear main gear of a 44th BS B-47E to the Wing's Chaplain. (*Paul Frye*)

of ejection, which found its way to many later aircraft including the B-52. The ejection sequence for the navigator was as follows.

He would place his feet in the seat-mounted stirrups, then quickly drop the visor on his helmet, pull the "Green Apple" of his emergency oxygen bottle to start it flowing, raise the left and right seat leg braces, and then pull a triangular D ring. Next, the ejection seat hatch was blown and supposedly flipped into the airstream below, pulling on a cable attached to the seat that continued the sequence, firing the ejection seat downward. After being ejected clear of the aircraft, the navigator would separate from his seat with the automatic release of his lap belt and a forceful push from the seat "butt kicker", which propelled him forward with his survival pack attached. Freefall ensued until the navigator passed through 15,000 feet, at which altitude his parachute would automatically deploy.

Phase II and III Modifications

Phase II and III modifications were relatively minor, encompassing only two significant upgrades. The first of these was a Phase III electronic countermeasures (ECM) package for jamming enemy air intercept radars. The addition of Phase III ECM was noticeable externally by the bulge on the bottom of the fuselage just ahead of the empenage. An AN/ALT-6 jammer was in this bulge. Some later E models utilized the subsequent AN/ALT-7, -8, -9, and -16 versions of this jammer in varying combinations. The ECM equipment was used along with an AN/ALE-1 Chaff Dispenser. Together they provided the only other means of defense for the Stratojet than its 20-millimeter cannon and in earlier years its speed. B-47E-IIIs were enhanced with updated three-phase constant-speed electrical alternators. Dependable electric current was vital to the B-47 as it powered much of the aircraft's mission equipment.

landing and go-around pattern (typical)

(AIRSPEEDS BASED ON AN AVERAGE LANDING GROSS WEIGHT OF 110,000 LB.)*

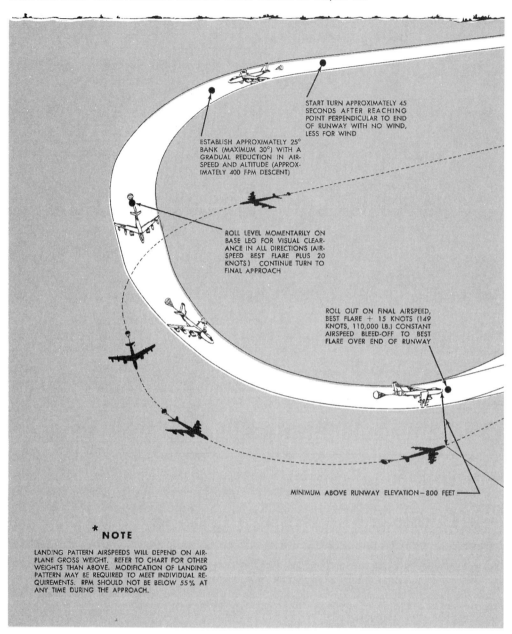

START TURN APPROXIMATELY 45 SECONDS AFTER REACHING POINT PERPENDICULAR TO END OF RUNWAY WITH NO WIND, LESS FOR WIND

ESTABLISH APPROXIMATELY 25° BANK (MAXIMUM 30°) WITH A GRADUAL REDUCTION IN AIRSPEED AND ALTITUDE (APPROXIMATELY 400 FPM DESCENT)

ROLL LEVEL MOMENTARILY ON BASE LEG FOR VISUAL CLEARANCE IN ALL DIRECTIONS (AIRSPEED BEST FLARE PLUS 20 KNOTS) CONTINUE TURN TO FINAL APPROACH

ROLL OUT ON FINAL AIRSPEED, BEST FLARE + 15 KNOTS (149 KNOTS, 110,000 LB.) CONSTANT AIRSPEED BLEED-OFF TO BEST FLARE OVER END OF RUNWAY

MINIMUM ABOVE RUNWAY ELEVATION — 800 FEET

*NOTE

LANDING PATTERN AIRSPEEDS WILL DEPEND ON AIRPLANE GROSS WEIGHT. REFER TO CHART FOR OTHER WEIGHTS THAN ABOVE. MODIFICATION OF LANDING PATTERN MAY BE REQUIRED TO MEET INDIVIDUAL REQUIREMENTS. RPM SHOULD NOT BE BELOW 55% AT ANY TIME DURING THE APPROACH.

Landing and go-around pattern for the B-47E. The B-47 was a "numbers" airplane in many senses. Strict adherence to flying the numbers kept crews safe. (*Jack Wright*)

Phase IV Modifications

Some of the most significant changes to the configuration of the B-47E were included in Phase IV modifications. Initially, internal fuel capacity fell in the E model to 14,610 gallons. Presumably, this was deemed practical because of the aircraft's air-to-air refueling capability and because of the extra fuel it was possible to carry externally in the two 1,780-gallon wing tanks introduced on the B-47B. However, with the B-47E-IV came the return of efforts aimed at bettering the Stratojet's unrefueled range. Structural improvements and a much strengthened new set of landing gear enabled the B-47E-IV to roll down the runway at a

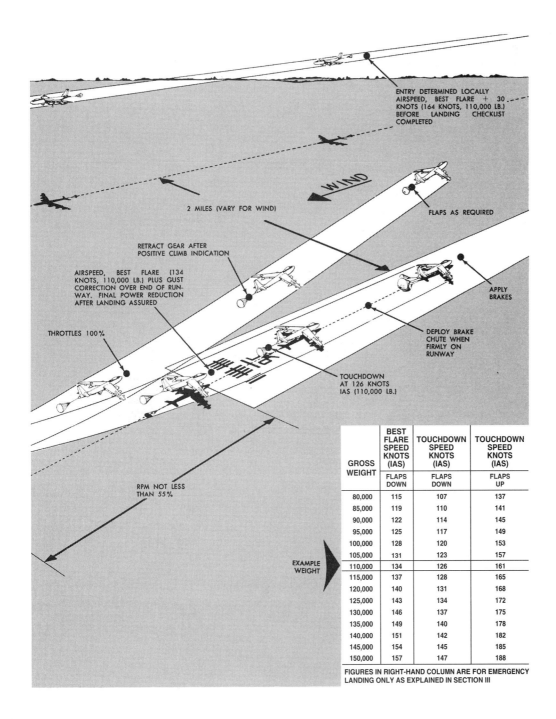

ENTRY DETERMINED LOCALLY AIRSPEED, BEST FLARE + 30 KNOTS (164 KNOTS, 110,000 LB.) BEFORE LANDING CHECKLIST COMPLETED

WIND

2 MILES (VARY FOR WIND)

FLAPS AS REQUIRED

RETRACT GEAR AFTER POSITIVE CLIMB INDICATION

AIRSPEED, BEST FLARE (134 KNOTS, 110,000 LB.) PLUS GUST CORRECTION OVER END OF RUN- WAY. FINAL POWER REDUCTION AFTER LANDING ASSURED

APPLY BRAKES

THROTTLES 100%

DEPLOY BRAKE CHUTE WHEN FIRMLY ON RUNWAY

TOUCHDOWN AT 126 KNOTS IAS (110,000 LB.)

RPM NOT LESS THAN 55%

EXAMPLE WEIGHT

GROSS WEIGHT	BEST FLARE SPEED KNOTS (IAS) FLAPS DOWN	TOUCHDOWN SPEED KNOTS (IAS) FLAPS DOWN	TOUCHDOWN SPEED KNOTS (IAS) FLAPS UP
80,000	115	107	137
85,000	119	110	141
90,000	122	114	145
95,000	125	117	149
100,000	128	120	153
105,000	131	123	157
110,000	134	126	161
115,000	137	128	165
120,000	140	131	168
125,000	143	134	172
130,000	146	137	175
135,000	149	140	178
140,000	151	142	182
145,000	154	145	185
150,000	157	147	188

FIGURES IN RIGHT-HAND COLUMN ARE FOR EMERGENCY LANDING ONLY AS EXPLAINED IN SECTION III

maximum takeoff weight approaching 230,000 pounds, 28,000 more than the B-47B. Most of the increased carrying capacity was devoted to additional internal fuel, which extended the Stratojet's combat radius to 2,050 nautical miles. While the increase in range was help- ful, the added heft of 28,000 pounds was not. Once again, more weight offset the increased power of GE's latest J47s. The extra capacity not consumed by fuel was given over to an increased bomb-toting ability. A 25,000-pound mix of bombs, conventional or nuclear, could be packed into the B-47E-IV's bomb bay. The advances in technology which lead to less cumbersome nuclear weapons allowed the majority of B-47Es to feature a short bomb bay configuration. A related measure undertaken to increase the safety of the B-47 in the event of a nuclear strike was the spraying of the undersurfaces and lower portions of its fuselage with glossy white paint designed to reflect the heat radiation from a nuclear blast.

An early B-47E in flight. Changes between the E and the B-47B are visible in the form of 20-millimeter cannon tail guns and the lack of internal ATO in the rear fuselage above the rear gear. (*National Archives*)

A close-up view of the B-47E's twin 20-millimeter cannons. These were controlled by the copilot who fired them using the A-5 FCS. (*National Archives*)

The slick polished nose of a B-47E. Forward of the radome is the navigator's escape hatch through which he would eject in the event of an in-flight emergency. Further back is the crew ladder and entry door. The glazed glass panels of the XB-47, B-47A, and B-47B were all but gone on the E model. (*National Archives*)

Pinecastle AFB, Orlando, Florida was home to B-47s of the 321st and 19th BWs, and with its flavor of the tropics must have been a desirable location for Stratojet crew assignments. (*National Archives*)

Note the long wire antenna running from aft of the canopy to the tail on this B-47E. Also visible are the aircraft's short bomb bay doors. (*National Archives*)

One of the most important upgrades to come from Phase IV modifications was the inclusion of the updated MA-7 Bombing-Navigation System (BNS). The BNS was central to the Stratojet's mission, and more B-47 navigators had experience with the MA-7A than any of the previous K systems. The MA-7 was state of the art, a new generation of equipment which evolved from B-29 and B-36 radar systems. Over 50 major components, weighing in

Noticeable on this B-47E are the 1,780-gallon external fuel tanks which were introduced on the B-47B. (*National Archives*)

An armament technician at Boeing-Wichita inspects a brand new B-47E's "stinger," its radar controlled 20-millimeter cannons. (*National Archives*)

excess of 1,600 pounds, along with more than 360 vacuum tubes embedded in numerous amplifiers made up the system. The heart of the MA-7 was the AN/APS-64, an excellent bombing radar capable of precision radar return definition. Integrated with analog navigation/bombing computers, it featured a large 10-inch Plan Position Indicator (PPI) radar scope and an 0–15 radar camera which recorded radar images for later interpretation. Computer and radar operating controls were located to the left of the navigator's seat. Each control was individually shaped and could be identified by touch. Depending on altitude, the AN/APS-64 had a maximum range of over 240 miles; the higher the altitude, the better the line of sight. Two modes were available, beacon and mapping. In beacon mode the radar interrogated incoming beacons from other bombers or tankers. Its main purpose was to aid in formation join-ups or rendezvous with tankers. Two rendezvous radar control panels to the left of the navigator mounted APN-69 and APN-76 transceivers, which transmitted a reply to the other aircraft for station keeping.

In radar mapping mode, the radar scope could be configured for high- or low-altitude navigation. Reportedly, the radar was good at low level, even with a line of site as little as 35 miles. A valuable feature was the capability to select a pencil beam or fan beam. The pencil beam concentrated radar power in a very narrow beam for a focused radar return. The fan beam reflected in a wide spread from the aircraft to the ground directly below. The pencil beam was used at low level much like terrain following radar. It could be locked onto the frame or gyros of the aircraft and stabilized for pitch and roll. The radar picture would indi-

Another B-47E slides easily through the skies above the Midwest. (*Boeing, courtesy of Jack Wright*)

A B-47E at Pinecastle AFB in 1956. (*National Archives*)

navigator's station—forward

1. Emergency Escape Hatch Release Handle	22. Polar Navigation Control
2. Variac	23. Bombsight Cover Heater Switch
3. Navigation-Computer Primary Control	24. Navigator's Work Table
4. Radar Scope	25. Relief Containers
5. Line of Sight Control	26. Bombsight Cover Retracting Crank
6. Navigator's Instrument Panel	27. Microphone Foot Switch
7. Azimuth and Sighting Angle Indicator	28. Navigator's Data Drawer
8. Navigator's Clock	29. Navigator's Tracking Control
9. Automatic Pilot Ready Light	30. Downward Ejection Hatch Pressure Gage
10. Periscopic Sextant Mount	31. Navigator's Upper Air Outlet
11. Oxygen Regulator Panel	32. Lockpin Inspection Window
12. Radar Power Monitoring Panel	33. Periscopic Bombsight
13. True Heading Transmitter	34. Downward Ejection Hatch
14. Bomb Intervalometer Panel	35. Radar Control Unit
15. Emergency Bomb Release	36. Radar Pressurizing Control Panel
16. Lighting Control Panel	37. Radar Transmitter Control Panel
17. Aerial Camera Intervalometer	38. Rendezvous Radar Control Panel (AN/APN-69) ①▶
18. Manual Bomb Release Switch	39. Rendezvous Radar Control Panel (AN/APN-76)
19. Aerial Camera Control Panel	40. Radar Camera Control Panels
20. Interphone or Intercom Panel	41. T-35 (IFCT) or SWK-2/A24T-1 Control Panel
21. Bomb Control Panel	42. T-19 (IFCT) or T-249 Control Panel

①▶ (Some Airplanes)

The B-47E Navigator's station. (*Jack Wright*)

cate the terrain (mountains or flatland) at the level of the aircraft, giving the crew confidence to navigate safely on a low-level trainer or Radar Bomb Scoring (RBS) sortie. A sensitivity time control unit allowed for the customization of radar returns on the display.

In conjunction with the AN/APS-64 there was an optical bomb site integrated with the mission computers so that its crosshairs would track in exactly the same way as the radar crosshairs. The optical sight combined left and right eyepieces with the optics in the right eyepiece. In the left eyepiece was an additional cathode ray tube radar display that showed a line-of-sight radar image to a selected target. The telescopic bombsight-moveable lens was located on the nose, protected by a rotating cover and slightly offset to allow room for the refueling receptacle. Inside it were lenses of different magnifications. The optics were infrequently used, however, their most common applications were in tactical bombing, picking out tanks or trucks on the ground. Clouds and darkness limited the optical bombsight's usefulness, and the AN/APS-64 was the main component by which the aircraft navigated or bombed.

The BNS computer navigation controls to the right of the navigator allowed him to calculate and display the coordinates of the aircraft's present position, the position of the radar crosshairs, wind components, and magnetic variation. In the aircraft commander's cockpit, the pilot's position data indicator (PDI) integrated with the BNS and displayed the direction to the bomb release loci or turning point and the "time to go." A red light on the PDI illuminated when the computer was switched to bombing mode and extinguished at the end of the countdown to "bombs away." The PDI could also be used for navigator-directed radar approaches to runways. A tracking handle allowed the navigator's right hand to control the radar crosshairs or optical crosshairs and the option to crank in the coordinates to a specific target. The handle also governed the autopilot, allowing directional control when in bomb function to steer the aircraft to the bomb release loci. A slew button on the tracking handle made possible a faster movement of the crosshairs if required. When the computer function switch was placed in track or bomb mode, the crosshairs could be

Men at work. The three-man crew of a B-47E ride together. The navigator stands in the aisle to the left of the pilot and copilot positions. (*Boeing, courtesy of Jack Wright*)

placed on a target or navigation point and observed for drift off the selected spot. If movement occurred, the tracking handle would be used to put the crosshairs back on the nav point, correcting the wind solution. The computers calculated ground speed and drift while indicators on the navigator's instrument panel displayed the sighting angle and azimuth of the crosshairs.

Target identification was confirmed by entering offset aiming points into the computers. Once the target had been confirmed and the time to go had counted down, the navigator could open the bomb bay doors manually with the flip of a switch or choose to let the computers open them automatically. Bomb control panels were to the either side of the navigator. The right-hand panel featured bomb door controls, individual bomb indicator lights, and the manual bomb release or "pickle switch." Approaching the target on a typical radar bomb scoring (RBS) run the navigator would turn on a tone at 20 seconds to go and confirm it, saying, "tone on." At 10 seconds to go the navigator would note the aircraft's true heading for scoring purposes and bomb damage assessment. Ten seconds later, the copilot would announce "bombs away!" over the UHF radio to the RBS site.

Training Missions

Most training missions comprised several objectives, which might include air-to-air refueling, radar navigation, an RBS run at low level, a climb to altitude for a 1½-hour celestial navigation leg, another RBS run at high altitude, and a radar penetration to landing. RBS sorties were frequent and allowed SAC to economically judge a crew's skill in bombing targets without actually dropping live ordinance. RBS sites relied on a combination of radar and radio to determine the accuracy of a simulated bomb run. Bomb scoring radars locked

Making an approach. A B-47E with its slipway refueling door open makes a careful approach to a KC-97. (*Boeing, courtesy of Jack Wright*)

onto an approaching Stratojet and tracked it automatically on a plotting board. Inbound, the aircraft would begin transmitting a tone shortly before (normally 20 seconds) electronic bomb release. The tone ceased at the instant of weapons release, allowing RBS personnel to fix the B-47's position relative to the target. A series of calculations accounting for variables such as, altitude, wind, distance, direction, aircraft speed, heading, and the characteristics of the weapon being simulated yielded the exact position a bomb or salvo of bombs would have impacted. Crews received scores based on the accuracy of their bomb runs. Interestingly, pilots could also be graded on the combat turn they executed pulling off the target.

Mission performance evaluations were a way of life for Stratojet crews and support personnel. SAC's philosophy of readiness dictated that all units be prepared for action at any moment. To this end, every bomb wing in the command was required to undergo an annual Operational Readiness Inspection (ORI). No notice of when these inspections might take place was given. Consequently, B-47 bomb wings had to constantly be ready to demonstrate their ability to go to war. The units were graded from top to bottom and evaluated in every operational category, from maintenance and supply to the performance of the flight crews on RBS runs.

Banking steeply, this crew shows the agility of their B-47E. (*National Archives*)

Plugged in! A B-47E takes on a load of fuel from a KC-97 tanker in the skies over Florida in 1954. The navigator's face is clearly visible peering up through the plexiglass frame atop the nose. (*National Archives*)

The constant pressure of preparing and training for the mission was a bit hard to withstand at times but B-47 crews knew they would be ready if ever called upon to actually execute it. SAC's War Plan assigned each B-47 crew a specific target. Effectively, however, each crew had two target assignments. First-priority targets were assigned to deployed Stratojet crews standing alert at overseas bases because of their proximity. The same B-47 crews on alert within the continental United States would have comprised the second wave of the strike force and would have required precisely coordinated tanker support. Accordingly, these aircraft were assigned secondary targets.

Aerial Refueling

Aerial refueling was a part of nearly every Stratojet sortie. Over its career the B-47 took on fuel from two different Boeing tankers a generation apart. The most common refueler of the B-47E was the KC-97. Millions of pounds of JP-4 jet fuel were dispensed through KC-97 booms to Stratojets. It is interesting to note, however, that the B-47 could also burn regular avgas. In fact, there were many instances in which KC-97s running low on JP-4 while pumping gas to a thirsty Stratojet would simply transfer 115/145 avgas from their own internal fuel tanks to the waiting B-47s, pumping it right on top of the JP-4 already off-loaded. Naturally, the Stratojet did not burn the avgas quite as efficiently as JP-4 and suffered some performance degradation but it was feasible and safe. As has been illustrated, the speed mismatch between the two airplanes was large. This caused a number of problems. The Stratojet struggled to stay in the air behind the KC-97, riding the edge of a stall. Meanwhile, the crews of KC-97s had everything wide open to maintain a high enough speed for the operation to be completed. Other performance limitations lead to costly inefficiencies. B-47Es operating at 35,000 feet would have to lose nearly half of their altitude to hook up with a KC-97. The four Pratt & Whitney R-4360s powering the tanker lost efficiency at higher altitudes, and on a good day, with its big radial engines firewalled, a KC-97 could barely struggle up to 18,000 feet. The fuel burned by B-47s in descending for refueling and making a careful rendezvous effectively halved the net gain of a full load of fuel pumped through the tanker's boom. Refueling behind the new KC-135 Stratotanker brought with it many gains in efficiency and safety. B-47s no longer had to descend to get gas and could take on a full load of fuel without penalty. Refueling in formation was much easier because of the greater speed of the KC-135. The only difficulty was that the KC-135 was actually a little too fast. B-47 pilots had their throttles pushed all the way forward to catch the fast Stratotankers and had to be careful as well of a "bow wave" of turbulent air behind the KC-135. Generally though, refueling from KC-135s was much safer.

Crew Requirements

Crew coordination was essential to effective employment of the Stratojet, especially when conducting strikes. Each member of the three-man crew was fully engaged

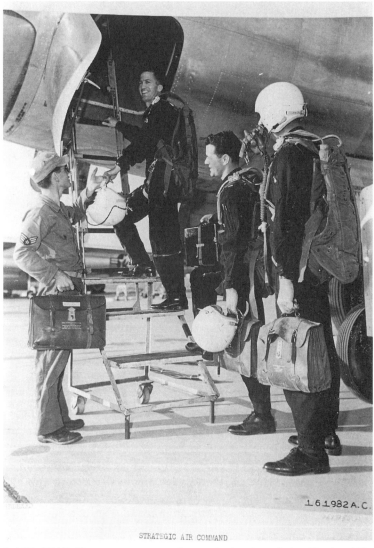

A 44th BW B-47 crew mounts up for a mission from Lake Charles AFB, Louisiana in 1956. (*National Archives*)

A streamlined B-47E flies over a flat expanse of the Midwest. (*Boeing, courtesy of Jack Wright*)

throughout a mission, none more so than the navigator. Much of the discussion of the B-47 has been told from the pilot's perspective but the role Stratojet navigators played was also crucial. They could make or break a crew in several areas of mission performance, and many crews achieved "select crew" status or won spot promotions based in part on the work of their navigators. Navigating a B-47 was serious business, demanding great concentration.

> **Low-altitude missions were a constant challenge. I always tried to stay 25 to 30 miles ahead of the aircraft and to think in advance about what was coming up next. There were so many variables that made navigating at low level difficult, the way you tuned the radar or the weather for instance. Humidity could be a problem. I've flown low-level missions over Missouri down in the hills when there was a temperature inversion and you couldn't see anything. It just blanked out your radar scope. Those were times when you needed to know where you were within a half mile all the time, especially when you where doing 7 miles a minute on the deck. You were so busy with checklists, calling them off with the copilot and trying to get the right headings. In the summertime out here in West Kansas, you'd bounce around at low level so bad that you had to put your knees around the radar scope to hold it so you could read it!**
>
> **JACK WRIGHT, NAVIGATOR, 40TH BW & 96TH BW, 1959–1965**

Not only was the navigator's work challenging, it was performed in surroundings and under conditions that were frequently less than comfortable.

I would usually get airsick on missions that required low-level navigation and bombing for 2 or 3 hours in the heat of summer. Flying as low as 500 feet above the terrain and being bounced around by thermals while inside a cramped, dim nose did not make me happy. Working from the table on the right to the radar scope on the left caused vertigo. Add a hot helmet with a mask plus lots of perspiration and stress and you get the heaves. If you were lucky, you remembered to bring a barf bag or extra cardboard burp cups. The flexible wings managed to soak up some of the turbulence, but it was hard to bounce in unison with the radar scope. I always made sure that I had something to eat prior to a flight that would be easy to handle and keep down. Peppermint Patties helped.

ANDY LABOSKY, NAVIGATOR, 376TH BW & 96TH SAW, 1960–1966

Missions could last up to 15 hours and were a test of endurance in many ways. Crew members fought fatigue and the discomfort of hours spent on a hard ejection seat. For the most part, crews were busy the entire flight, especially the navigators, and on many occasions the flight lunches stowed within easy reach went uneaten. Still, there was the odd moment to observe one's surroundings and wonder at them.

To the front of the Nav station and slightly to the right was the air-refueling receptacle. During aerial refueling the noise level in the nose rose quite a bit when the receptacle's slipway doors opened. When the tanker's boom was inserted into the receptacle, the fuel intake was quite noticeable. The fuel pipe curved down and then was routed under the navigator's table to make its way to the fuel tanks. I always thought it was odd that an ashtray was welded onto the pipe! A flange connecting two sections of the curved pipe was just in front of the navigator, and sometimes the seal would leak during air refueling, spraying me and the scope which operated at 15,000 volts!

ANDY LABOSKY

The Stratojet shows off its maneuverability in this series of photos as it pitches up into the LABS toss bombing maneuver and rolls inverted. Pilots claim it could execute such maneuvers well with little airframe stress. (*National Archives*)

. . . And over the top. (*National Archives*)

Evolving Tactics

By the mid-1950s the advanced Stratojet had begun to lose its edge in certain areas. It was still the finest operational strategic bomber in service, but new weapons systems applying the latest in technological innovations were beginning to catch up with the B-47. One area of great advancement was in the capabilities of surface-to-air missiles (SAM). Their range and accuracy were better than ever and the Soviets took full advantage of this, fielding newer, more capable SAMs. Consequently, flying high and fast no longer guaranteed safety from SAMs, and the B-47 was vulnerable. Responding to the threat, the Air Force decided to change its techniques for penetrating enemy airspace. The B-47 would now go in low, evading enemy defenses by staying below their radar coverage until just before reaching the target. In accordance with this new tactic, two methods of weapons delivery were introduced.

The first of these was known as the LABS (Low Altitude Bombing System) maneuver. This was essentially a toss-bombing technique. An attacking B-47 would make a low-level penetration, approaching its target at high speed. Several miles from the target, the aircraft would pull up into a steep climb. On the way up, a bomb would be released at a predetermined angle and was literally lobbed at the target.

. . . And back around, leveling out. (*National Archives*)

The maneuver was much like an Immelman Turn and had the advantage of permitting a quick departure from the target area. The altitude gained in the half loop would be converted into airspeed as the B-47 made a quick descent back to low level to "get out of Dodge." One of the architects of this maneuver was pilot George Birdsong, who, by 1955, was the Chief of Tactics at SAC headquarters.

We had three Wings qualified to do the LABS maneuver, the 306th, 310th, and 22nd. If we would have had to go to war at that time the LABS may have been the only way we would've gotten bombs in on Moscow and several other targets. It worked because if you went in at around 200 to 300 feet the Russian radars couldn't pick you up until you were about 25 miles out. If you were at 35,000 feet they could pick you up 250 miles out. We would run in to the target at about 425 knots and then haul back on the yoke pulling positive G's all the way up and over. We dropped the bomb at about a 45 degree, nose up angle and it would freefall for about 8 miles in front of you with remarkable accuracy, we were surprised to find out. Then we'd pull over the top and half-roll out on a 180 degree heading. It really was the best way of delivering a bomb until we got later weapons called *lay down* weapons.

GEORGE BIRDSONG

The second method, sometimes called the *pop up method,* was similar, involving the same low-level penetration technique of the LABS maneuver. It differed in that the bomber would make a rapid climb to medium altitude—popping up—a short distance from the target for its bomb run. After dropping the bomb, the B-47 would turn hard away from the target and begin a rapid descent to make its escape. One hundred, twenty-five B-47Es were modified for low-level operations in 1955.

The crews of the 306th, 310th, and 22nd BWs received low-level training in a program which got under way in 1956 called *Hairclipper.* Early in 1958 six B-47Es were lost in accidents during low-level bomb runs. A brief suspension in low-level operations was called to allow investigators to look into the causes of these accidents. It was determined that airframe fatigue was the source of the problem, specifically, failures of the main wing-fuselage fittings called *milk bottle bolts.* Subsequently, a massive modification program known as *Project Milk Bottle* was launched with the aim of eliminating these failures, and virtually the entire B-47 fleet was inspected and strengthened with the replacement of worn or cracked milk bottle bolts. The inspections and modifications were carried out in five locations, including Boeing-Wichita, Douglas-Tulsa, Lockheed-Marietta, Oklahoma City Air Materiel Command at Tinker AFB, and McClellan Air Materiel Command at McClellan AFB, Sacramento, California. By September 1958 Stratojet bomb wings were resuming operations at low level, executing the pop up maneuver in favor of the LABS. This was mostly as a result of the new generation of lay down weapons which had since been developed and which were more suited to delivery by the pop up method.

Unfortunately, several commentators have drawn the conclusion that the accidents of 1958 and the subsequent Milk Bottle program were the direct result of airframe overstress caused by the LABS maneuver. Evidence suggests otherwise. Though only three combat wings were qualified to perform the LABS maneuver, the majority of SAC bomb wings engaged in low-level operations from the mid-1950s onward. Low-level RBS runs and navigation at low altitude were routine. Pilots who flew the LABS report that the aircraft executed it easily with no excess airframe stress. The LABS maneuver itself was

A ground crew look over a B-47E. (*National Archives*)

SAC B-47 ground crews were some of the best and heartiest in the business. Foul weather did not keep Stratojets from flying. In this photo, a B-47E at a northern air force base is being readied for flight. Note the hose thrust through the crew entry door to pump heat into the cold bomber. (*National Archives*)

not the problem. Rather, it was the pounding that the B-47 took at low altitude, especially in turbulence, that caused fatigue. Signs of fatigue were present in aircraft which never flew the LABS maneuver but which did fly in low-level operations. This was part of the logic behind SAC's order that the entire B-47 fleet be put through the Milk Bottle program.

SAC's commitment to low-level operations was initially quite large. Plans originally called for 1,000 B-47s to be modified for low-level flying, which would've entailed the installation of absolute altimeters, terrain avoidance equipment, and doppler radar to most of the fleet. In the end, a number of factors including testing delays, the Milk Bottle program, and a lack of funds which lead to the phase out of several bomb wings caused SAC to scale down its requirements to only 500 aircraft. Despite a new sense of urgency, stemming from an assessment declaring that all B-47s would be obsolete by 1963 if not adapted for low-level missions, the modification program was further trimmed back to just 350 Stratojets. Once again, the reason was a lack of funds.

Crew Support

Equally as dedicated and skilled as the flight crews who operated the Stratojet were the ground crews who maintained them. The crew chiefs and mechanics on the flightline at B-47-equipped SAC bases around the world were the real "owners" of the aircraft. They took immense pride in their work and were essential to B-47 operations. Many of these men enjoyed long careers in the Air Force and had experience maintaining several generations of aircraft. By and large they claim that the Stratojet was reliable and easy to maintain.

I made the transition from working on B-29s to B-47Es. The B-47E was a dream to work on and, generally, easy to maintain. When they first arrived they did have a few bugs which needed to be worked out, though. Fuel booster pumps had a habit of leaking and a few aircraft were lost because of this problem before a true fix was established. Flight crews always felt their pucker strings tighten when told to count and time the number of drops per minute of fuel leakage. It was like strapping a bomb onto your butt and gauging how large an explosion might ensue! Engine "choo choo" was another problem. The B-47's engines had a habit of "choo-chooing," running unevenly at times, sounding like a steam locomotive. It was caused by fuel pressure fluctuation and fuel flow surges. It really messed up the flight crew's power settings when it happened and you never knew when it would happen.

It was fixed a short time later by adding an accumulator to the engine to even out the fuel pressure and fuel flow from the fuel tank booster pumps. The hardest problem to overcome was the Bomb/Nav equipment. We did greatly increase its reliability but I'm not sure all the problems were ever completely overcome.

DON CAREY, CREW CHIEF, 40TH BW, 44TH & 45TH BS, 1953–1957

B-47 ground crews worked hard, often around the clock to keep their aircraft flying. Crew Chiefs and their ground crew would walk out to an aircraft 3 to 5 hours prior to takeoff to complete their own preflight. Inspections were thorough, including checks of the fuel boost pumps, checks for hydraulic leaks in all areas, checks of tire pressures, landing gear strut extension, the aircraft battery (the B-47's engines could be started on its own battery power without use of auxiliary power units), navigation lights, landing lights, and instrument panel lights. Fuel tank readings were taken with dipsticks to com-

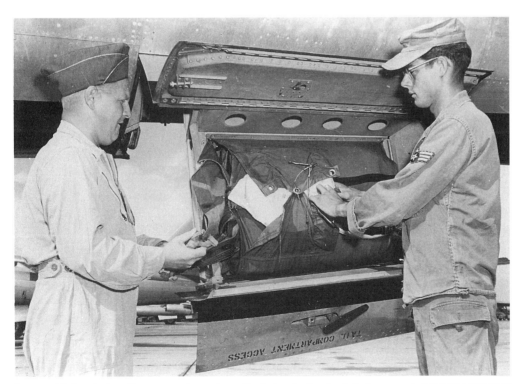

A crew chief and aircraft commander make a final check of a drag chute, assuring themselves that the chute is packed correctly. This was part of every preflight. Ground crews did exacting and essential work, laboring long hours to keep B-47s flying and ready. (*National Archives*)

A SAC sentry stands guard over B-47Es at Barksdale AFB in 1954. (*National Archives*)

pare against the fuel gauges. The aircraft was serviced with water and alcohol in the summertime for augmented takeoffs. Drag chutes and brake chutes were inspected to ensure proper installation. A multitude of additional checks would be completed by the time the flight crew arrived at its assigned aircraft approximately 1½ hours before takeoff. The flight crew would then do its own thorough exterior inspection, connecting the drag and brake chutes while the Crew Chief followed closing access panels. The flight crew then boarded their aircraft to continue the interior preflight while the crew chief communicated with them via headset. After visually checking all of the lights and control surfaces, removing chocks, landing gear locks, and closing the main entrance door, the crew chief would assist in sequenced engine starts, confirming rotation and combustion of each engine. When all was ready the flight crew would give a "thumbs up" and the crew chief would direct the aircraft from its parking spot. Indicative of the confidence which many flight crews had in their support personnel was their willingness to let the ground crews to do most of the preflight, making only the briefest of exterior inspections, focusing on their interior checks.

100th BW B-47Es from Portsmouth AFB pass over the USS Constitution in salute to King Saud of Saudi Arabia as he arrived for an official state visit to the United States. (*National Archives*)

The flight crews had a lot of faith in the crew chiefs and ground crews as at times all they would do was connect the drag and tell you to close the panels and we would all go have a cup of coffee. Then there were times when we knew they didn't want to fly. They didn't come right out and tell you but you had that feeling, and they would abort takeoff for a fire warning light because they knew it would take time to fix or that we'd find nothing.

PAUL FRYE, CREW CHIEF, 40TH BW, 44TH BS, 1954–1960

Operational Changes

Operational changes in the employment of the B-47 at home and overseas took hold in the mid- and late-1950s. Prior to 1957 there had been no scheduled alert system for Stratojet crews. Bomb Wings were on 24-hour alert, meaning that they could launch strikes within 24 hours of the order; however, no crews or aircraft were kept in a continuous state of vigilance, ready to scramble at the sound of a klaxon. But, that year SAC adopted a One-Third alert policy. Like the MITO program, this was in response to the increased threat from Soviet ICBMs. The doctrine called for all SAC Bomb Wings to keep one-third or approximately 15 of their 45 aircraft on ground alert, fully armed and fueled, ready to launch immediately. The command conducted three separate evaluations of the *alert force concept,* operations *Try Out, Watch Tower,* and *Fresh Approach,* to assure itself that the One-Third alert would be feasible. In October 1957 General Thomas S. Power, SAC's new commander, directed that all Bomb Wings begin operations in this manner. The policy brought with it a number of administrative and organizational changes that lead to the addition of an extra squadron to each Wing.

THREE THINGS ON THE TAIL	— CHECK
THINGS THAT MOVE ON THOSE THINGS	— CHECK
TWO LITTLE PEDAL THINGS	— MOVE
STEERING WHEEL MOVES	— CHECK
HORN BUTTON – PUT IN POCKET	— GOT IT
GAS GAGE NEEDLE ON BIG NUMBER	— YUP
NAV – GIVE US A HACK	— HACK – THURSDAY

A standard preflight checklist, tongue-in-cheek, sketched by B-47 navigator, Andy Labosky. (*Andrew Labosky*)

Changes were under way at SAC bases abroad as well. The number of B-47s deployed outside the continental United States had grown considerably since the first Stratojet deployment made by the 306th BW to Fairford, England in 1953. In addition to Fairford, there were now several more bases on English soil and elsewhere in Europe, North Africa, and the Pacific. The expansion was largely due to a change in overseas deployment policy instituted in the summer of 1957. As U.S. bases adopted a One-Third alert structure to enhance deterrence and capability, forward bases did likewise, ending the operational rotation method of 90-day TDYs for every Bomb Wing. A new alert system code named *Reflex Action* was launched in late 1957 and early 1958. The differences in this method of deployment enabled forward based units to achieve and maintain a much higher level of combat readiness. Reflex Action deployments were not Wing-sized deployments. Rather, Bomb Wings would rotate only a portion of their aircraft and crew to a forward base for 3-week periods, usually keeping 15 to 20 aircraft deployed. In a sense, overseas bases became a second home for Stratojet flight crews and support personnel. A particular crew or group of crews would fly from their home base, stateside, to a forward base assigned to their Wing. Upon arrival the aircraft would go on alert immediately. The aircraft would be turned around and put in "cocked" configuration—fueled and armed—ready to launch within 15 minutes. The aircraft would remain on alert, cocked and ready, for the entire three weeks. The crews themselves were on a different schedule. A constant rotation of crews and air-craft was the routine. As one crew finished a 3-week tour and readied themselves for the

An alert crew of the 307th BW at Lincoln AFB, Nebraska scrambles to their aircraft during a practice alert in 1960. (*National Archives*)

trip back, a fresh crew and aircraft arrived to take their place. Consequently, there were always more crews than airplanes. This allowed the pilots and navigators to have up to a week off during the 21-day period.

We'd put an airplane on alert for a week, have a week off, then be on alert for the third week. It was a welcome getaway from the atmosphere back in the States where they were constantly looking over your shoulder. It was tough on your family, but when you were on alert at your home base you were being hassled all the time, tested and retested, having ORI's (Operational Readiness Inspections), etc. So it was a pleasure to get out of there. Plus, we knew we'd have 5 days off between alert periods and we could go anywhere we wanted to. If we were in England we'd jump on a C-54 and it would take us wherever we wanted to go, anywhere on the continent. There were some really wild stories that resulted from those 5-day breaks. We'd come back and go on alert to rest up from the "R & R"! It was a pleasurable change, especially in the summer.

ANDY LABOSKY

A formation of Stratojets fly by in a demonstration during the National Airshow in Philadelphia in 1955. (*National Archives*)

Deployed B-47Es of SAC's Eighth Air Force line up, awaiting a predawn takeoff from the snow-covered runways of Thule AFB, Greenland. (*National Archives*)

One of the 22nd BW's B-47Es, part of a trio parked on the ramp at Yakota AFB, Japan after completing a 6,700-mile nonstop flight from March AFB, CA. The three Stratojets under the command of Major General W. C. Sweeney, commander of SAC's 5th Air Force, completed the flight in 15 hours with three inflight refuelings. The flight took place on June 25, 1954. (*National Archives*)

Three Stratojets of the 22nd BW from March AFB, CA on the flight line at Yakota AFB, Japan along with B-29s of SAC's 3rd Air Division and an early C-124. Note the two stars on the nose of General Sweeney's B-47E in the foreground. (*National Archives*)

Later Deployments

Overseas deployments of the Stratojet reached their zenith in 1958 when B-47Es making Reflex Action deployments could be found at bases including RAF Fairford, Brize-Norton, Greenham Common, Upper Heyford, Bruntingthorpe, and Chelveston in the United Kingdom—Sidi Slimane, Ben Guerrir, and Nouasseur in Morocco—and at Moron, Torrejon, and Zaragoza in Spain. B-47s could be seen on the other side of the world as well, although deployments in the Pacific were never as numerous as those in Europe. Beginning in 1956, B-47Es from a succession of Bomb Wings made Air Mail deployments, these being essentially the same as Reflex Action, to Anderson AFB, Guam, and to Eilson and Elmendorf AFBs in Alaska. Stratojets were seen at other far-flung locations around the world, but these visits were made mostly by reconnaissance versions of the aircraft. The only other location B-47Es formally visited was Yakota Air Base, Japan. In June 1954 three Stratojets from the 22nd BW set a record for the longest point-to-point flight by the B-47 to that point in time, flying nonstop from March AFB to Yakota. They completed the 6,700-mile flight in less than 15 hours.

The period between 1955 and 1960 was a time of ironic contrasts for the B-47E. Expansion of the B-47 force was explosive, peaking in December 1956 by which time over 1,300 examples were in service. In the same month, SAC gave a resounding demonstration of its ability to strike quickly and massively when, in response to the Suez Crisis, over 1,000 B-47s flew nonstop simulated combat missions over the Arctic and continental United States, averaging 8,000 miles per sortie. Yet within less than a year, the phase out of the Stratojet began as the 93rd BW at Castle AFB turned in their B-47Es for the new B-52. Moreover, just 9 months earlier, in February 1957, the last B-47E built (53-6244) was delivered to the 100th BW at Pease AFB. Simultaneously, the aircraft was setting records and winning awards. On

Stratojets from several bomb wings on the ramp at Walker AFB, New Mexico for the 6th Annual Bombing and Navigation Meet. SAC's B-47 crews pitted themselves against each other and the best crews from B-29, B-50, and B-36 Wings. (*National Archives*)

The "City of Orlando II" was part of the a one-two finish, along with its sister B-47E, the "City of Winter Park," for the 321st BW in SAC's 1957 Bombing and Navigation competition. (*National Archives*)

The "City of Winter Park," a B-47E of the 321st BW along with its sister 321st B-47E, "City of Orlando II," won the Bombing and Navigation awards in the 1957 SAC competition (*National Archives*)

January 25, 1957 a B-47E flew from March AFB, California to Hanscom AFB, Massachusetts in 3 hours, 47 minutes, averaging 710 mph. In August 1957, a 321st BW Stratojet flew 11,450 miles nonstop from Andersen AFB, Guam to Sidi Slimane Air Base, Morocco in a record 22 hours, 50 minutes. Showing its mettle in competition, the B-47E finally bested SAC's very experienced B-36 units in 1955. The 320th BW broke through to win the command's hotly contested annual Bombing and Navigation competition that year. The Fairchild Trophy and the spoils of victory went to B-47E units three more times, with the 321st BW, 306th BW and 307th BW taking home the trophy in 1957, 1958, and 1959. Finally, in November 1959 a B-47 from the Wright Air Development Center stayed aloft for an amazing 3 days, 8 hours, 36 minutes, flying more than 39,000 miles.

As the 1960s dawned the sun began to set on B-47 operations. However, operational exigencies lead to a number of contradictory decisions being made about the future of Stratojet, and for a time the B-47 seemed to be going in two different directions at once. While a single Wing was retired during the course of 1959, 27 additional Stratojet squadrons joined the force between September 1958 and May 1959, completing a reorganization of Bomb Wings which resulted from the One-Third alert. Plans to phase out the Stratojet were accelerated by President John F. Kennedy in the Spring of 1961 at the recommendation of his Secretary of Defense, Robert McNamara, who felt a greater emphasis should be placed on strategic missile systems. At the same time, another operational change in the B-47 alert force was taking place. Increasing tensions between the United States and the Soviet Union prompted the President to order that SAC maintain fully half of its force on alert, armed, fueled, and ready to launch. Instituting the 50 percent alert was a major undertaking for SAC, which involved another reorganization of its force structure and, ironically, the large batch of additional squadrons added little more than a year before were gone by October 1961. Several Stratojet Wings were phased out during 1960 but the topsy-turvy drawdown

Row upon row of obsolete Stratojets are lined up in the 2704th Air Force Aircraft Storage and Disposition Area at Davis-Monthan AFB, Arizona in November 1963. (*National Archives*)

Large formations of B-47s such as this one passing in review with F-84Fs above must've been an awesome sight and sound! (*National Archives*)

A view of a B-47E, head-on. (*National Archives*)

process was interrupted for a period, first by the Berlin Crisis of 1961 and then by the Cuban Missile Crisis in 1962, during which B-47s were dispersed to widely scattered civilian and non-SAC military airfields. Only one B-47 Wing was disestablished in 1962. Between 1963 and 1964 ten Wings were retired and as the next year began just eight Stratojet equipped Wings remained. A number of the Wings phasing out the B-47 transitioned to the B-52, and in a couple of cases the new B-58 Hustler. Interestingly, there was an effort to export some of the B-47s retiring from service during this period. In November 1963 the USAF flew three B-47Es as demonstrators for Australian flight crews in *Project Australia,* a proposal which would have leased two squadrons of B-47Es to the RAAF to help fill an interim period before delivery of their F/B-111s.

By 1965 hundreds of B-47s had made their way to the Davis-Monthan AFB, Arizona to be put in long-term storage and, ultimately, destroyed. The final axe blow to the B-47 program came in October 1965 with the commencement of *Project Fast Fly.* The effort was aimed at bringing the phase-out to a quick conclusion and the last two active SAC B-47 Wings took their B-47Es to Davis Monthan AFB in February 1966. This was not the B-47E's last hurrah, however. A number of E models served well into the future as flying test beds, and the final Stratojet flight in history took place nearly 20 years later on June 17, 1986 when a restored B-47E (52-0166) was flown from the Naval Weapons Center at China Lake, California to Castle AFB, California to be put on static display at the Castle Air Museum.

Specifications for the B-47E are as follows:

Performance

Maximum speed   606 mph/527 knots at
    16,300 ft 557 mph/491 knots at 35,000 ft
Cruise speed   500 mph/434 knots
Stall speed   175 mph/152 knots
Service ceiling   33,100 ft
Combat ceiling   40,500 ft
Initial climb rate   2,430 ft/min
Max climb rate   4,660 ft/min
Combat radius   2013 mi w/10,845-lb bombload
Maximum range   4,035 mi w/16,318 gal fuel load
Takeoff ground run   10,400 ft, 7,350 ft w/ATO

Dimensions

Wingspan   116 ft
Length   107 ft, 0 in
Height   27 ft, 11 in
Wing area   1,428 sq ft

Armament

Two 20-mm M24A1 cannons

Bombload

25,000 lb

Weights

Empty   79,074 lb
Combat   133,030 lb
Gross   198,180 lb
Maximum takeoff   230,000 lb

Crew

Three: pilot, copilot/radio
    operator/gunner, bombardier/
    navigator

Serial Numbers

B-47E   51-2357 through 51-5257, 51-7019 through 51-15812, 51-17368 through 52-409, 52—411 through 52-620, 52-1406 through 52-3373, 53-1819 through 53-2315, 53-2317 through 53-2409, 53-2411 through 53-4244, 53-6193 through 53-6244

WB-47E   51-2357, 2445, 2358, 2360, 2362, 2363, 2366, 2369, 2373, 2375, 2380, 2383, 2385, 2387, 2390, 2396, 2397, 2402, 2406, 2408, 2412-2415, 2417, 2420, 2427, 2435, 5218, 5257, 7019-7083, 7021, 7046, 7049, 7058, 7063, 7066,

YDB-47E   51-5214- 5257, 5219, 5220

JB-47E   52-389

DB-47E   53-2345, 2346

EB-47L   52-305,

EB-47L/ETB-47E   53-4207 through 53-4244, 53-6193 through 53-6244; completed as either model

B-47E Variants and Conversions

As the classic Stratojet, the B-47E was by far the most numerous of the type. Not surprisingly, improvizations on its design were numerous as well. In fact, more variants were spun off from the B-47E than the B-47B. Not only were there more E model derivatives, more examples of these variants were produced, one version yielding more than 200 aircraft.

Almost as quickly as it recognized the B-47B's potential for taking on new missions and special projects, the Air Force realized that the B-47E could meet these needs even more capably. Indeed, some of the earliest B-47B variants were completed as E model variants or were modified right along with B-47Es being adapted to new missions. Moreover, the range of missions undertaken by B-47E variants was even wider than those tackled by versions of the B-47B. The Air Force pushed the Stratojet airframe to the limit, assigning it duties never envisioned by the Boeing engineers who designed it. B-47Es in a host of configurations performed missions including aerial reconnaissance, electronic intelligence gathering, electronic countermeasures, weapons system testing, crew training, communications relay, and drone/target duties. Building on the success of earlier models B-47E variants brilliantly demonstrated the versatility of the Stratojet.

EB-47E

By the mid-1950s advances in technology were being applied across a wide spectrum of offensive and defensive military capabilities. As we have already seen, this affected the Stratojet in a number of ways. In addition to the hardware physically deployed by both the East and West, the two sides were preparing to fight on an increasingly sophisticated electronic battlefield. Early warning and detection systems were now a fact of life and like the West, the Soviet Union ringed its borders with a network of radar systems to search for any sign of intrusion.

To successfully reach their assigned targets, a strike force of B-47s would have to penetrate these defenses. Numerous reconnaissance versions of the Stratojet were used to map and identify the various components of the Soviet defense network, marking areas of weak-

The 1000th Stratojet, a B-47E flies in formation with the 1,001st Stratojet, an RB-47E. (*Boeing, courtesy of Jack Wright*)

ness and gathering intelligence to help evade and defeat the latest antiair systems. Still, it was clear that striking B-47s would have to run a gauntlet of radar-guided fighters and surface-to-air missiles to reach their targets. To help them punch through, the Air Force planned to disrupt the Soviet defenses using electronic countermeasures (ECM).

A relatively new science, ECM was used to jam or confuse radar and communications signals across a wide range of frequencies. The initial and most common application of ECM to Stratojets was a series of jammers and chaff dispensers installed in all B-47Bs and Es. The jammers and chaff offered a modest self-protection capability but SAC wanted more. In 1955, between 60 and 100 B-47E-1s of the 303rd and 509th Bomb Wings were fitted with additional ECM equipment carried in two external pods mounted on the fuselage on both sides of the bomb bay. The pods, known as *Tee Town pods,* each carried four AN/ALT-6B jammers, and these Stratojets were known as *Tee Town B-47s.* The Tee Town B-47s retained their bomb-carrying ability along with the new jammers, and so would have been among the aircraft most likely to reach their targets.

It was obvious, however, that more powerful and effective jammers were needed to give the entire force a better chance of getting through. This lead to the development of Stratojets dedicated solely to ECM. The first of these were B-47Es fitted with a special ECM suite. This Phase IV ECM package consisted of 16 jammers, initially AN/ALT-6Bs, carried in a removable cradle mounted in the bomb bay. ACE-1 chaff dispensers and APS-54 warning receivers completed the package. The aircraft were designated EB-47Es and their reconfigured bomb bay gave rise to the name by which the aircraft were known, *Blue Cradle.* The Phase IV Blue Cradle EB-47Es were operated by the 376th BW at Lockbourne AFB, Ohio.

Their sister bomb wing at Lockbourne, the 301st BW, operated a more advanced version of the EB-47E, which differed considerably from the Phase IV EB-47Es flown by the 376th. Phase V EB-47Es were modified to carry a pressurized capsule in their bomb bay. Inside this compartment, two Electronic Warfare Officers (EWOs) sat side by side at two work stations operating 13 different jammers including nine AN/ALT-6Bs (high band, GHI band, and D band), one AN/ALT-7, and one AN/ALT-8 (EF band). Whereas Phase IV EB-47E crews could jam a range of radar and communications frequencies, Phase V EB-47E EWOs could focus their ECM, selecting specific signals from threat radars or communications networks to be jammed and rendered ineffective. Phase V EB-47Es received an update approximately 2 years into their service, featuring even more capable jammers including the QRC-49 (EF band), QRC-65 (VHF communications jammer), QRC-95, QRC-96, QRC-139, ALT-13 (D through I band), and the ALT-15.

The tactics employed by EB-47E crews were designed to confuse and complicate the Soviet electronic picture, diverting attention away from the main elements of the B-47 strike force on its way into Russian airspace. They required guts and a certain amount of fatalism as EB-47E crews knew they had little chance of survival. If the klaxons sounded they would take off with the rest of the bombers standing alert. Typically, the EB-47Es and bombers would penetrate enemy airspace simultaneously. The bombers would go in at low level, streaking toward their targets at high speed, using either the LABS or pop up maneuver when in the target area. Meanwhile, the Phase IV and V EB-47Es would knife through Soviet airspace at high altitude, zooming toward Moscow and Leningrad, jamming and laying chaff. They were "magnets" for enemy radars, obscuring their picture and drawing attention away from the bombers at low level. Of course, the very jamming signals they sent out could be homed in on. At their high altitude they would be quite vulnerable and Soviet defenders would make them high-priority targets. The crews recognized this and knew they were for practical purposes, expendable. Still, if they could disrupt the defenses for long enough, the bombers would have a better chance of striking their targets. Approximately 40 B-47Es were modified as Phase IV and V EB-47Es.

Two more B-47Es were modified as EB-47Es but these aircraft never served with the USAF. Only one Stratojet ever served outside the United States, the previously described CL-52, which flew with the RCAF as a test bed for the Orenda Iroquois engine. Otherwise, the entire fleet of Stratojets flew with the USAF with three exceptions. These B-47s were flown by the U.S. Navy. Two of the three Navy Stratojets were former 376th BW B-47Es loaned by the Air Force in 1965. USAF Stratojets 52-410 and 52-412 became Navy 24100 and 24120, respectively. These aircraft were modified at the Douglas-Tulsa plant and were flown by Douglas flight crews in support of the USN's Fleet Electronic Warfare Support Group (FEWSG). These Navy EB-47Es were operated chiefly from NAS Point Mugu on the West Coast and NAS Oceana on the East coast, testing naval ECM equipment and acting as "Orange-Air" aggressors in fleet exercises.

Each of the aircraft received black "US NAVY" block identification letters, replacing their USAF identification. They were otherwise unadorned with the exception of their new serial numbers. Each aircraft carried a variety of ECM gear with external ECM pods mounted on the wing auxiliary fuel tank pylons. Other antennas and chaff dispensers were installed as well. Ironically, 24100 and 24120 were the last operational Stratojets, serving with the USN long after the last USAF B-47s were retired. In fact, the last operational flight of a B-47 took place on December 20, 1977 when LtCol. David Hall and copilot Robert Tuttle ferried Navy 24100 from Douglas-Tulsa to Pease AFB, New Hampshire where it was put on static display. This aircraft is now in storage at Ellsworth AFB, South Dakota. 24120 is now on display at Dyess AFB, Texas.

There was one further version of the EB-47E, the EB-47E(TT). This model, however, was so unique that it bears inclusion in the Special Variants section of this book.

RB-47E

While it is true that the B-47 never dropped a bomb in anger or was directly involved in any armed conflict, some Stratojets did go in harm's way on a regular basis. On several occasions B-47s engaged in the dangerous business of aerial reconnaissance did see combat.

The RB-47E was the most numerous of the dedicated reconnaissance Stratojets. Two hundred and forty were built for photographic reconnaissance. Begun in 1951 as RB-47Bs, these aircraft incorporated many of the upgrades, including engines and defensive armament, due for the B-47E. The first of the new-build RB-47s rolled off the line at Boeing-Wichita in 1954. By this time the aircraft had so much in common with the E model that its designation was changed to RB-47E. The new recon Stratojets were easy to distinguish from their bomber brethren. The nose of the bomber was extended by 34 inches, giving the RB-47E a more streamlined look. The extra space accommodated an air-conditioned compartment, containing four types of cameras. An O-15 radar camera and forward oblique camera were used for low-altitude work, while K-17 trimetrogon cameras and two 36-inch focal length, K-36 target cameras were used to cover larger areas of interest. In all, up to eleven cameras could be employed. The sensitive black-and-white-film cameras were operated by the navigator, who took the new designation, photographer/navigator. The RB-47E had a nighttime photo recon capability as well, with the use of onboard photoflash bombs and cartridges. As a dedicated reconnaissance platform the aircraft had no bombing equipment but did retain its inflight refueling capability. The RB-47E was somewhat longer-legged than the bomber with an increased maximum internal fuel load of 18,405 gallons. The RB-47E retained as well the defensive armament of the E model, a capability it would have occasion to use.

B-47s operated at all hours of the day but none more so than the reconnaissance versions of the Stratojet. Here, a 55th SRW RB-47E returns to Forbes AFB, Kansas after a night mission. (*National Archives*)

A formation of RB-47Es cruises along at 35,000 feet in May 1956. This view offers a good look at the pylon-mounted, podded J47, GE engines. Also note the wing flex and the tiny vortex generators three quarters of the way down the wing. (*National Archives*)

On May 8, 1954 my copilot Carl Holt, navigator Vance Heavlin, and I overflew approximately 700 miles of the northwest Soviet Union to cover nine airfield targets during the middle of the day to determine if the Bison bomber or MiG-17 fighter had been deployed. Our mission proved the bomber had not been deployed (which my crew didn't find out until years later), but we proved first hand that the MiG-17 was in service because at least 12 of them chased us, six at a time. One got a lucky hit through our left wing and into the fuselage forward main tank, exercising the self-sealing feature and knocking out our intercom, which is very unhandy in a tandem airplane! The aircraft flew fine, though. We still had all six engines running and made a successful emergency air refueling to get the film safely back to the United Kingdom. The three of us have a soft spot in our hearts for the RB-47E!

HAROLD "HAL" AUSTIN, PILOT, 91ST SRW 1953-1956

This was not the last time recon versions of the Stratojet would be attacked, but it did prove the ruggedness and survivability of the aircraft. Future B-47 recon crews relied on the aircraft's fine performance as well, even in the face of improving Soviet defenses.

A typical mission in the RB-47E might be comprised of an initial navigation leg, a rendezvous with a KC-97 for gas, then on to photograph a number of assigned targets, many at low level (often as low as 500 feet), then a final navigation leg and return to base. Photo recon sorties were long and fatiguing with durations of between 8 and 16 hours. Mission assignments varied, including very specific targets or wide geographical areas. For instance, RB-47Es were used to continue work begun by RB-45Cs for the U.S. Army Map Service, mapping Europe and the United Kingdom.

An RB-47E tops off from a KC-97 tanker. (*Boeing, courtesy of Jack Wright*)

The RB-47E's service life was rather brief, however. The aircraft began to be phased out in the fall of 1957. Several factors shortened their career, including the development of superior aircraft designed specifically for photo reconnaissance, such as the high-altitude U-2/TR-l, and the shift to collection of electronic intelligence (ELINT) in the Stratojet recon community. Though most of the photo recon Stratojets were retired from service prior to 1960, exotic ELINT B-47s continued to serve throughout the 1960s.

Serial numbers: 51-5258 to 51-5276, 51-15821 to 51-15853, 52-685 to 52-825, 52-3374 to 52-4264.

YDB-47E/DB-47E

The YDB-47E and DB-47E were part of the Air Force/Bell development program for the Bell Gam-63 Rascal air-to-surface guided missile program.

Two B-47Es were converted to YDB-47Es for simulations tests with the Rascal as the program encountered difficulties already described in the section on the YDB-47B/DB-47B. By 1955, the program had been greatly reduced in scope. Plans to modify significant numbers of E models as Rascal carriers were abandoned, and ultimately only 74 GAM-63 capable YDB/DB-47Bs were completed. Two additional B-47Es were modified as DB-47Es, and it was one of these aircraft which actually made the first successful launch of the Rascal in July 1955. These DB-47Es later served as Drone Directors with the 3205th Drone Group at Eglin AFB, Florida.

Altogether, just these four YDB/DB-47Es were produced and the entire program was canceled in 1958.

Deployed to Alaska. The RB-47Es of the 90th SRW are shown parked on the ramp at Eilson AFB, Alaska in March 1956. The RB-47E had a purely photographic reconnaissance capability. (*National Archives*)

ETB-47E

In the late 1950s a number of B-47Es were equipped with a fourth crew seat for an instructor to serve as training aircraft, replacing older TB-47Bs. These aircraft were designated ETB-47Es and served until the early 1960s.

JB-47E

The designation JB-47E was given to a small number of B-47Es modified for special test work. These flying test bed duties were normally temporary, and the aircraft involved were returned to standard B-47E configuration once their projects were completed. An example is JB-47E-110-BW (53-2280), which is now on display at the Air Force Museum. This aircraft was used by the Wright Air Development Center at Wright-Patterson AFB from 1967 to 1969 as a test bed for "fly-by-wire" primary flight controls. It was the first USAF aircraft to employ the system and successfully proved the concept.

QB-47E

Carrying on the well-established practice of using the Stratojet for a variety of systems test programs, 14 RB-47Es were converted as drone aircraft between 1959 and 1960. They were among the most colorful Stratojets ever to fly, featuring high-visibility day-glo paint.

Known as QB-47Es, the aircraft were operated by the 3205th Drone Director Group at Eglin AFB as part of the effort to develop surface-to-air missiles like Boeing's own IM-99 Bomarc. The QB-47Es were essentially targets and went through an expensive modification program to allow them to be flown by remote control. Conversion began with the installation of radio control equipment, which enabled operators on the ground in spe-

This DB-47E is 79 seconds away from a hot firing of the Bell Gam-63 Rascal over the Holloman Air Development Center Test Range, New Mexico. (*National Archives*)

cially equipped trucks to launch and recover the unmanned QB-47Es while drone director aircraft including DB-47Es took over positive control once the drones were airborne. To aid in recovery, an arrester tail hook was mounted behind the rear main gear and was deployed on landing just after touchdown. To help gather data during missile testing, special pods attached to the auxiliary fuel tank pylons on each wing carried six 16-millimeter movie cameras. In the event that positive control of an unmanned QB-47E was lost, self-destruct

One of the two DB-47Es sits on the ramp at Holloman AFB, New Mexico with a Bell Gam-63 Rascal missile mounted on its starboard fuselage in October 1956. (*National Archives*)

Gulp! RB-47Es, like every other Stratojet, were thirsty aircraft. Inflight refueling was a part of life for any B-47 crew, but the aircraft was reportedly good in formation and not overly difficult to refuel in flight. (*National Archives*)

explosives were carried onboard which could be detonated by timers or by the drone director aircraft.

Operational doctrine for the QB-47E called for most of its unmanned or "Nullo" flights to take place over the Gulf of Mexico where missiles could be fired at the drones safely with no fear of incidents involving the civilian population. But after totaling up the expense involved in making these Stratojets drones, the Air Force decided not to make them missile fodder, at least initially. Consequently, missiles launched against these flying targets were programmed to miss, albeit by a slim enough margin, to allow conclusive findings about their accuracy to be obtained. At this point, the curious reader may be wondering if any of the missiles ever accidentally struck their targets. Apparently, they did and on at least one occasion a QB-47E was shot down by a Bomarc which scored a direct hit. Eventually, most of the QB-47Es were destroyed during missile tests and by the early 1970s just two of the drones remained in storage at Davis-Monthan AFB. It appears that these last two were specially designated as JQB-47Es and were involved in test work with the Hughes AIM-47, a long-range air-to-air missile. This weapon was originally intended for use by Lockheed's YF-12 (the forerunner of the SR-71).

JTB-47E

Two JTB-47Es were assigned to the Rome Air Research and Development Center (RADC) at Griffiss AFB, New York during 1960. Other sources have referred to these aircraft as JRB-47Es but RADC records and the Air Force Research Library Information Directorate archives indicate that they were in fact JTB-47Es.

The nature of the work these aircraft were engaged in is not well documented but it appears that they flew in support of classified ECM programs. Outfitted with various ECM

These RB-47Es at Eglin AFB in August 1959 have been modified by Lockheed to serve as QB-47E drones. (*National Archives*)

equipment, evidence suggests they may have been used to test the readiness of CONUS early warning radar systems operated by NORAD (North American Aerospace Defence Command).

JRB-47E

While records indicate that no "JRB"-47E ever served at the RADC, it seems that such an aircraft did exist. Several photos show this oddball Stratojet. It appears to be an extensively modified RB-47E, featuring thin tubular antennas along its lower fuselage, extending from just in front of the forward main gear aft. In addition, it carried a large antenna atop the fuselage near the leading edge of the wings, a nose probe, and a deep nonstandard radome. The antennas along the lower fuselage are not unlike those used on the very special variant EB-47E(TT)s for telemetry intelligence missions. The antenna atop the fuselage may provide some clue as well, its location suggesting that it was meant to collect signals from above. Discussions by the author with knowledgeable members of the Stratojet ELINT community lead to the conclusion that this aircraft may have been used to prove the technology later utilized by the EB-47E(TT)s in collecting telemetry data from Soviet missile and satellite launches. The "J" designation may reasonably be assumed to have been attached to this special RB-47E as it was applied to most test aircraft. As far as is known only a single JRB-47E existed.

YB-47J

As previously detailed, utilization of the Stratojet as a flying test bed was quite common. Yet another Stratojet to serve in this capacity was the YB-47J. A single B-47E was given this designation in the early 1950s and was modified to test the MA-2 radar bombing/navigation system being developed for the B-52. Further research suggests, however, that as many as

A QB-47E makes a manned/remote control landing at Eglin AFB. These aircraft were used as drones in missile tests and incorporated ECM, scoring, and telemetry systems for use in evaluation of area defense systems and airborne countermeasures. An F-80 drone director aircraft accompanies the QB-47E on landing. (*National Archives*)

10 to 17 B-47Es may have flown with the MA-2 during flight testing of the system. Still, only one aircraft is known to have had the designation YB-47J. The trials flown with the MA-2 helped speed the introduction of the Stratofortress.

EB-47L

In the early 1960s the Air Force and Navy realized the vulnerability of ground-based command and control networks to nuclear attack. To allow control of the U.S. nuclear triad of strategic bombers, ballistic missile submarines (SSBN) and intercontinental ballistic missiles (ICBM) to be maintained in the event of such an attack, both services instituted the use of airborne command and control platforms which could maintain communications if ground-based networks were decimated. In 1961, the Air Force began the "Looking Glass" worldwide airborne command post mission using the EC-135. Simultaneously, the USN stood up an airborne command post squadron of EC-130Qs to control its SSBNs.

To aid in the relay of communications between these airborne command posts, ground stations and other aircraft, the Air Force modified as many as 35 B-47Es in 1963 with AN/ARC-89 communications relay transceivers. Designated as EB-47Ls, these aircraft worked in conjunction with EC-135s and ground stations, providing line of sight relay of general purpose communications. EB-47Ls were stationed around the continental United States, serving with a number of different bomb wings. Typical missions were 8 hours in duration and provided little excitement for the EB-47L's three-man crews. The crews simply turned on their transceivers, orbited at a certain position, and allowed messages to funnel through their aircraft, re-aiming and retransmitting them to different users. The EB-47L was an interim measure in airborne command, augmenting the range of the EC-135 communications to SAC aircraft and other users. New technologies were soon applied which

A ramp full of B-47Es of the 306th BW at MacDill AFB, Tampa, Florida. In the background is one of the Wing's KC-97s from the 306th ARS. (*National Archives*)

A QB-47E drone makes an approach. In the foreground is the terminal control truck which, employing special equipment, could take off and land the QB-47E. A range director truck or drone director aircraft controlled the QB-47E on the weapons range. (*National Archives*)

A close-up view of the countermeasures/chaff pod on a QB-47E drone. The countermeasures simulated conditions an enemy bomber or missile might create. (*National Archives*)

The single NB-47E. A B-47E loaned to the Navy and General Electric to act as a flying test bed for the GE TF34 turbofan which was under development for the new Lockheed S-3 Viking. (*Mark Natola*)

eliminated reliance upon line-of-sight communications and the EB-47Ls were phased out after two years of service in late 1965/early 1966.

RB-47K

Delivered in December 1955, the RB-47K was a new-build Stratojet developed from the RB-47E. In fact, the last 15 RB-47Es ordered by the Air Force (53-4265 to 53-4279) were completed as RB-47Ks and these were among the last Stratojets built. The RB-47K differed from the RB-47E by the addition of high-resolution, side-looking radars (SLAR) and air-sampling equipment. The K retained the photographic reconnaissance capability of the RB-47E but had a broader mission.

Assigned to the 338th Strategic Reconnaissance Squadron (SRS) of the 55th SRW, RB-47Ks had as their primary mission, meteorological data collection/weather reconnaissance. Photographic reconnaissance was their secondary mission. Stratojet crews newly assigned to the 55th usually flew RB-47Ks with the 338th SRS, performing the weather recon mission and gathering experience in the friendly skies over Canada before moving on to the Wing's more advanced ELINT Stratojets. Normally, two missions, *Weather Alpha* and *Weather Bravo*, were flown every day with special missions included from time to time. Each mission launched from Forbes AFB, Kansas, the first taking crews north to Sault Ste. Marie, then on to Hudson Bay, counterclockwise around the perimeter of the Bay and back to Forbes. The second took crews on the same route within the United States but traveled in an easterly direction over the Labrador Sea and to the north of Goose Bay, Canada. Flight time for these missions was approximately 12 hours.

The data collected by these sorties was used to develop War Plan meteorological condition predictions for weather around the Soviet Union. Sampling of radioactive fallout from

The last 15 RB-47Es were completed as RB-47Ks like this one. The Ks had both a weather and photo reconnaissance capability but their chief mission was to gather meteorological data on strategically significant areas around the Soviet Union. (*Mark Natola*)

foreign nuclear tests could also be taken. The aircraft's tools for gathering data included the SLAR and weather sensors called *dropsondes.*

Crews would release the dropsondes at designated points along the route of flight. The dropsonde dispensers held eight dropsondes and all eight were dropped on each mission. The dropsondes transmitted the data they collected back to a recording device at the navigator's station. I liken this device to a polygraph with four or more recording pens. The dropsondes were like microphones, taking in and transmitting weather data which was measured and recorded. The Ks were also equipped with cameras to photograph ground targets or survey a ground area. There were five cameras: a forward, left and right oblique cameras, and two vertical cameras of different focal lengths. With a few exceptions, more experienced crews from the other two 55th squadrons flew the special photo collection missions in the Ks over sites such as Russian ice islands and other less hostile areas. Photo missions flown by regular 338th crews were generally over disaster sites in the US but not on any regular basis.

JIM NELSON, NAVIGATOR, 55TH SRW, 1958–1964

The RB-47K served with the 55th SRW for 8 years, eventually being phased out in 1963. Its serial numbers were 53-4265 to 53-4279

NB-47E

The third Stratojet to see service with the U.S. Navy was the NB-47E. B-47E (32104) was bailed by the Air Force to the Navy to serve as a flying test bed for General Electric's TF-34-2 turbofan engine. The NB-47E was then given Navy markings and loaned to General Electric which was developing the TF-34 for the brand new Lockheed S-3 Viking. The aircraft flew as part of the program from 1969 to 1975. Interestingly, the NB-47E was piloted by a gentleman with test flight experience in another of the rarest Stratojets, the previously described YB-47F/KB-47G tanker version.

A WB-47E from MATS 9th Weather Reconnaissance Wing on one of hundreds of flights these Stratojets would make during a relatively brief career. (*National Archives*)

An RB-47K weather reconnaissance Stratojet of the 338th SRS, 55th SRW banks steeply. (*Mark Natola*)

> We tested the full envelope for the TF-34 engine. The advantage of using the Stratojet as a flying test bed was that we could test the engine in the same environment that the S-3 would operate in, altitude and speedwise. We mounted the engine on the left drop tank hardpoint. On takeoff, we throttled back the engines on the port side so that we could run the TF-34 at max power. It had no adverse effects on the way the B-47 flew. It was completely controllable.
>
> **CHARLES ANDERSON, NB-47E GE TEST PILOT, 1969–1975**

The NB-47E was one of the last four Stratojets flying in the 1970s. It made its final trip to the boneyard at Davis-Monthan AFB in the summer of 1975.

WB-47E

At the beginning of the 1960s, MATS Air Weather Service's 9th Weather Reconnaissance Wing was in need of a replacement for its aged, high-time WB-29 and WB-50 weather reconnaissance aircraft. In 1956 a single B-47B had been modified as a WB-47B to serve as a Hurricane Hunter with the 55th WRS at Hunter AFB, Georgia. But no more weather recon Stratojets had been procured.

Ultimately, the need grew critical for new and updated aircraft and the Air Force decided to improve upon the WB-47B. Thirty-four B-47Es were modified as WB-47Es at Lockheed's Marietta plant. The modifications included nose-mounted cameras which recorded cloud formations and a special sensor pod fitted in the bomb bay. The pod carried air sampling and data recording equipment to measure wind velocity and direction, temperature and pressure. The defensive A-5 FCS system was deleted and the 20-millimeter cannons were removed.

WB-47Es of the 53rd and 55th WRS deployed worldwide, gathering weather data. Often, they were launched in advance of deploying fighter squadrons to collect weather information along the routes the fighters would fly. Many WB-47Es built up considerable time on their airframes, logging hundreds of flights over a relatively brief period of service. The type began to be phased out in 1965 with the introduction of the WC-130 Hercules and WC-135 weather recon aircraft. The last WB-47E was retired on October 31, 1969. This aircraft was in fact the last operational Stratojet in the Air Force inventory.

Special Variants

While the Stratojet made its main contribution to the peace and security of the United States as a bomber, its value as an intelligence gathering tool should not be over-looked. The role played by Stratojets in collecting photographic intelligence, electronic intelligence, and telemetry intelligence (TELINT) was critical to our knowledge about Soviet offensive and defensive weapons systems and their order of battle. Bomber versions of the Stratojet and later B-52s, or for that matter any aircraft tasked with penetrating Russian airspace, would have relied heavily upon the information collected by the unique and special Stratojets of the 55th SRW. The manner in which these B-47s were operated and equipped differed widely enough from the bomber versions of the Stratojet that a separate look at these rare "special variants" is warranted.

RB-47H

The use of search radar/tracking systems by ground-based and airborne Soviet antiaircraft defense systems spread rapidly in the late 1940s and early 1950s. Realizing that SAC's force of strategic bombers, including the new Stratojet, would have to penetrate this formidable network to strike targets inside Russia, the Air Force began to look for ways of collecting information on these defensive systems to defeat them and give American bombers a better chance of getting through. The collection of data on Soviet radar systems got under way shortly after World War II. A variety of aircraft were modified by the USN and USAF to fly the ELINT mission including Navy PB4Y-2s and P2Vs, and Air Force RB-17s, EC-47s, RB-29s, and RB-50s. Even the newest of these ELINT aircraft, the RB-50, was outdated by the early 1950s and the search for a newer, higher-performance platform was begun.

Just as it had recognized that the Stratojet might be adapted to a number of other missions for which it was not originally designed, the Air Force felt that the B-47 might make a suitable ELINT aircraft. By early 1952, efforts were under way to design such a variant. When the first RB-47H was delivered to the 55th SRW in August of 1955 it was immediately apparent that this Stratojet was unlike other B-47s.

In physical appearance it most closely resembled the RB-47E, although there were a number of pronounced differences. Like the RB-47E, the RB-47H had a streamlined nose section which had been lengthened by almost three feet. But unlike the RB-47E, the RB-47H sported a large circular radome on the tip of its nose. The radome housed a forward-and-aft-looking APS-54 airborne radar detection system and an APD-4 receiver capable of detecting mid-frequency radar emissions.

More unusual bumps and protrusions were evident further aft along the fuselage. The area of the fuselage normally occupied by the bomb bay bulged, giving the aircraft a pregnant look. In place of the bomb bay a pressurized compartment which carried a crew of three EWOs (also known as "Ravens" or "Crows") was built into the fuselage. Up front, a pilot, co-pilot, and navigator occupied their normal positions, giving the RB-47H a crew of six.

The Crow compartment was packed with electronic eavesdropping equipment including radar receivers, oscilloscopes, amplifiers, recorders, and cameras. Inside, the three EWOs sat facing aft at three "Raven Stations." Just behind the Crow compartment on the underside of the fuselage was an ASQ-32 communications receiver antenna. Further aft, below the empennage on the bottom of the fuselage, was a large bulge housing two direction finder (DF) antennas which spun back to back, searching high and low frequencies. On the left and right sides of the fuselage in the same area were blisters for additional APD-4 mid-frequency receiver antennas. Two more APD-4 antennas could be found in fairings located on the aft ends of the outboard no. 1 and no. 6 engine pods.

In addition to its 20 mm tail guns, each RB-47H carried a battery of ECM jammers and chaff dispensers (many crews had occasion to deploy these). A total of 35 RB-47Hs were built. Three were completed as highly specialized ERB-47Hs. The last RB-47H was delivered in January of 1957. Though only 35 RB/ERB-47Hs were built, their contributions to intelligence were significant enough that a major upgrade for these aircraft, known as the "Mod 44" or "Silverking" update, was executed in 1961, giving these special variants newer,

The very first RB-47H (as denoted on its nose) rolls off the production line at Boeing's Wichita facility in 1955. (*National Archives*)

more capable sensors to keep pace with technology. The modifications consisted of new APR-17 receivers to replace the older APD-4s, the addition of an ASQ-32 communications receiver, video cameras, and other upgrades including the QRC-91 Airborne Warning/Detection system and the ALD-4 pod. The QRC-91 replaced the APS-54. Like the APS-54, the QRC-91 was meant to detect airborne intercept radars from hostile aircraft (i.e., Russian fighters). However, the QRC-91 had better sensitivity and could detect signals much farther away. It also gave a direction bearing to the hostile aircraft, a feature not available with the APS-54. The QRC-91 was mounted at the rear of the DF antenna bulge. The ASQ-32 receiver automatically searched for signals on three subcomponent receiver bands cov-

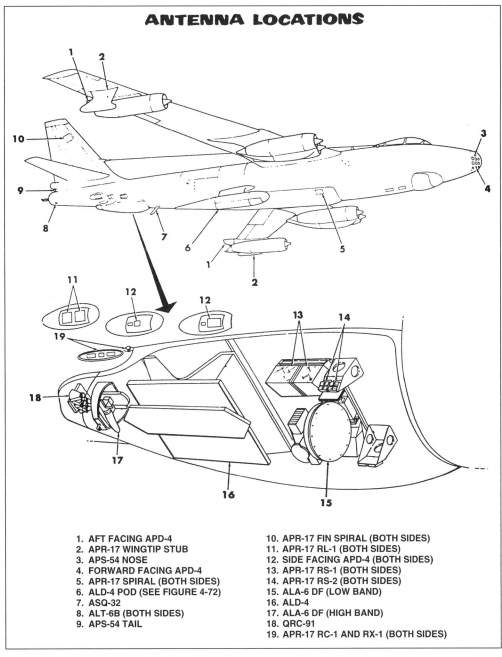

ANTENNA LOCATIONS

1. AFT FACING APD-4
2. APR-17 WINGTIP STUB
3. APS-54 NOSE
4. FORWARD FACING APD-4
5. APR-17 SPIRAL (BOTH SIDES)
6. ALD-4 POD (SEE FIGURE 4-72)
7. ASQ-32
8. ALT-6B (BOTH SIDES)
9. APS-54 TAIL
10. APR-17 FIN SPIRAL (BOTH SIDES)
11. APR-17 RL-1 (BOTH SIDES)
12. SIDE FACING APD-4 (BOTH SIDES)
13. APR-17 RS-1 (BOTH SIDES)
14. APR-17 RS-2 (BOTH SIDES)
15. ALA-6 DF (LOW BAND)
16. ALD-4
17. ALA-6 DF (HIGH BAND)
18. QRC-91
19. APR-17 RC-1 AND RX-1 (BOTH SIDES)

The complicated antenna array of the RB-47H is detailed here in a diagram from an RB-47H Dash-One flight manual. (*Dave Johnson*)

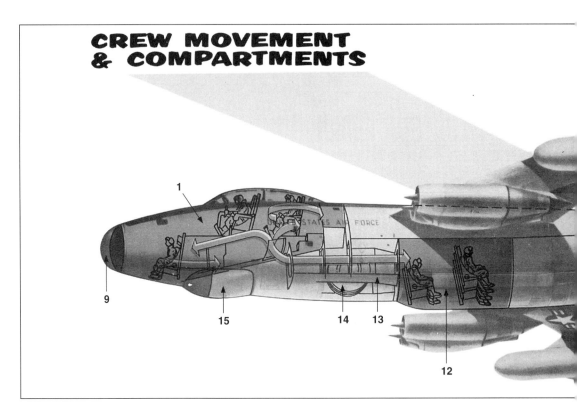

RB-47H crew movement and compartments from the aircraft's Dash-One flight manual. Note the Crow compartment and the narrow tube which Crows had to negotiate to reach it. (*Dave Johnson*)

ering different frequencies. Its blade antenna was located just behind the Crow compartment on the underside of the fuselage.

The ALD-4 was a very noticeable addition to the Silverking RB-47H. Mounted on a pylon extending from the lower right side of the fuselage below the trailing edge of the right wing, the ALD-4 gave Crows a more flexible way to collect data on high-priority signals from new or modified Soviet radars. Instead of having to spend entire missions monitoring one or two frequencies the ALD-4's very wide frequency automatic receivers allowed inspection of up to 10 high-priority signals at once, freeing Crows to search out specific signals on other frequencies. When a signal met a set of parameters programmed into the ALD-4 prior to the mission, the receiver would give an alarm, alerting the EWOs. They would then tune to the indicated frequency and manually probe the signal to gather as much technical data on it as possible. The signals were processed and recorded on digital, video, and analog tape systems.

Crows and The Mission

Crew coordination was essential in any Stratojet, but the premium placed on teamwork among the five- and six-man crews of ELINT Stratojets was especially high. The mission demanded it. RB-47H and ERB-47H crews from the 38th and 343rd SRSs played a calculated but dangerous game every time they flew an intelligence gathering sortie. To gain firsthand information on Soviet electronic defense systems, the crews literally had to provoke them into action. ELINT Stratojets had to present enough of a threat to Russian defenders that they would turn on their latest and greatest antiaircraft radar systems. In this way, the Crows could ferret out and identify new or modified threat radars. Signals from SAM sites, airborne intercept radars, early warning/tracking radars, and GCI datalink signals between ground stations and Soviet fighters were all processed, recorded, and analyzed.

Methods for stirring up activity varied. Often, crews would fly directly at sensitive, highly

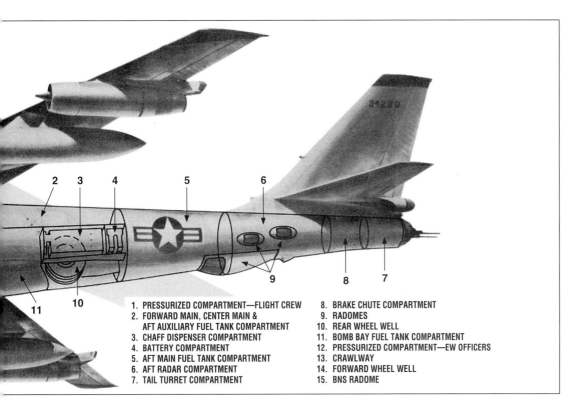

1. PRESSURIZED COMPARTMENT—FLIGHT CREW
2. FORWARD MAIN, CENTER MAIN &
   AFT AUXILIARY FUEL TANK COMPARTMENT
3. CHAFF DISPENSER COMPARTMENT
4. BATTERY COMPARTMENT
5. AFT MAIN FUEL TANK COMPARTMENT
6. AFT RADAR COMPARTMENT
7. TAIL TURRET COMPARTMENT
8. BRAKE CHUTE COMPARTMENT
9. RADOMES
10. REAR WHEEL WELL
11. BOMB BAY FUEL TANK COMPARTMENT
12. PRESSURIZED COMPARTMENT—EW OFFICERS
13. CRAWLWAY
14. FORWARD WHEEL WELL
15. BNS RADOME

defended areas along the Russian borders, turning away from them at the last possible moment to stay outside of Soviet airspace or to stay just outside the firing envelope of SAM sites. Another tactic was to descend rapidly after turning away from an area of interest, dropping down to very low altitude below radar coverage. Having shaken off Russian tracking systems the crew would then double back and re-enter the same area or another sensi-

A 55th SRW "Silverking" RB-47H. The most noticeable of the many modifications included in this upgrade program was the ALD-4 receiver pod which hangs from the pylon on the right side of the aft fuselage. (*Dave Johnson*)

tive sector unobserved. The pilots would then pull their aircraft up steeply, popping up to medium altitude. This normally stirred up the defenses considerably as the Soviets knew the intruder would be bearing back in on them at low altitude. As a consequence, many radars would be turned on. Immediately after popping up to medium altitude the Crows would begin processing and recording all of the ELINT data they could take in before being forced off or shot down.

The crews were used to having "company" on these missions as Soviet fighters of different types were frequently scrambled to intercept them. Maintaining the delicate balance between threatening the Soviets enough to gather the intelligence information needed without provoking hostile action was a tricky business. Pilots had to fly course headings precisely, relying on accurate position information from their navigators who constantly recalculated their position to be sure that they didn't inadvertently stray inside Russian airspace, giving Soviet fighters a legal opportunity to shoot them out of the sky. Of course, there were a few occasions on which crews intentionally penetrated Russian airspace but these were infrequent as the desire to cause international incidents was not high.

Close coordination between all crew members was essential but at the core of the mission were the Crows. The terms *Crow* and *Raven* were derived from *Project Raven,* an effort begun in World War II to exploit for the first time the new world of electronic warfare. The first combatants on the electronic battlefield were British radar observers who were given the nickname *Ravens.* Their American counterparts came to be known as *Crows* and both terms stuck. The environment in which these men worked aboard their special Stratojets was anything but hospitable. Crows earned the respect of other B-47 crew members not

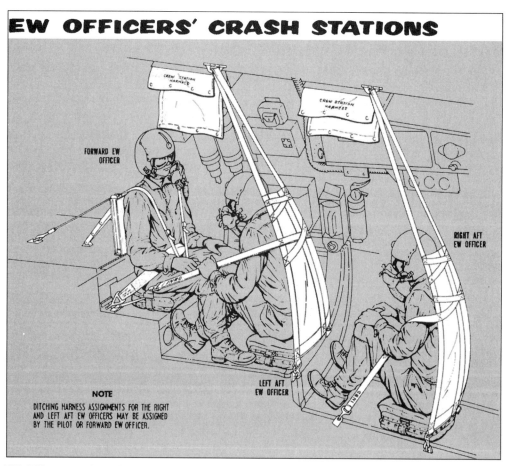

EW Officers crash stations aboard an RB-47H. This was where Crows were crammed together to endure the discomfort of takeoff before going to their compartment. (*Dave Johnson*)

only for their skills and expertise but for their willingness to do their job under less than ideal conditions. Just getting to their workplace was hazardous. For takeoff, the three EWOs were jammed into a small area on the metal floor to the left of the copilot's position. Two of the men sat in slings facing aft while the third sat on a step facing forward with the second Crow literally in his lap. Once a successful takeoff had been made and the aircraft was established at a steady rate of climb, the Crows struggled out of their take-off positions to a telescoping entrance ladder housed in a small cubicle above the aircraft's entrance hatch. Gingerly making their way down the ladder, they took great care to avoid stepping on the entrance hatch below which could not support their weight and could be knocked open with potentially serious consequences.

Next, the Crows had to gymnastically maneuver their way head first into a narrow crawl way which led back to the Crow compartment. Having successfully slithered through this tube, each man was forced to squeeze through a very small hatch at its end and into the compartment. Once inside, there was almost no room to move. Crawling on one's knees or on all fours was mandatory as the compartment measured less than 4 feet in height. The Raven Two and Three positions were side by side in the aft end of the space while the Raven One sat in the forward right corner. Each Crow strapped into a downward firing ejection seat. None of the seats had ejection hatches to fire through beneath them. If ejection was initiated, steel blades on a platform below the seats were to punch through the fuselage floor, making a hole through which the Crows would eject (tests of the seats showed that successful ejection would not have been possible; however, Crows were never informed of this).

Seated at their respective work stations, each Crow was surrounded by racks of receivers, radar scopes, analyzers, recorders, and cameras. The noise level inside the compartment was deafening because the thin aluminum skin of the fuselage which formed the compartment walls was not insulated. Moreover, the compartment itself sat between and just behind the inboard engines. The Crows invariably froze or boiled due to the poorly designed air conditioning/pressurization system. In addition, fuel tanks surrounded the compartment and, on

The ERB-47H was a highly specialized ELINT aircraft. Various bulges on its fuselage housed powerful antennas which the ERB used to gather electronic data on Soviet radar systems. (*National Archives*)

occasion, leaked fuel into it, causing the Crows to frantically shut down the hot electronic gear to avoid being vaporized in an explosion. Nevertheless, the Crows accepted these conditions as routine and endured them for 8 to 14 hours at a time, two or three times a week.

Despite the discomfort and distractions, Crows reveled in their work and did it with tremendous success. "On watch," Crows experienced a fast-paced continuum of intercepting large numbers of emissions, carefully tuning their receivers in search of new or unidentified signals which fit preprogrammed parameters. Once found, a specific signal was recorded and photographed and pinpointed by location.

An array of equipment available to the Ravens, One, Two and Three, could include:

APR-9 Receivers

APR-17 Receivers

APD-4 Receivers, Channels B and C

ALA-6 High Frequency DF Antennas

APA-74 Pulse Analyzer

KD-2 35 Millimeter Camera

ALA-5 Pulse Analyzer

QRC-91 Airborne Intercept Receiver

ASQ-32 Communications Receiver

APS-54 Airborne Intercept Receiver

The RB/ERB-47Hs of the 38th and 343rd SRSs were based at Forbes Air Force Base, Kansas along with the rest of the 55th's aircraft [RB-47Ks and EB-47E(TT)s]. However,

A good view of the underside of a "Silverking" RB-47H's fuselage, peppered with antennas. Aside from the very noticeable ALD-4 pod are APD-4 antennas in the nose and in blisters at the aft end of the fuselage. The large bulge on the after fuselage houses high- and low-band DF antennas. (*Dave Johnson*)

deployments for the hardworking ELINT crews were frequent. But the manner in which they deployed differed markedly from their Stratojet bomber brethren. Detachments were normally composed of one or two aircraft at most. Operating in as anonymous a fashion as possible, RB/ERB-47Hs found their way to the remotest corners of air force bases in Alaska, Greenland, England, Turkey, and Japan. Sequestered in hangars well away from the rest of the activity at these locations, the 55th crews left as small a footprint as possible and most of the base personnel had no idea what the ELINT Stratojets were doing. Base commanders exercised little if any authority over them and had scant knowledge at best of their purpose. Most missions were launched under the cover of darkness, recovering many hours later. Not counting complex briefs and debriefs, typical mission durations were 8 to 10 hours. If launching from bases which required inflight refueling to enable the crews to reach their areas of interest, missions could last as long as 16 hours. Crews making successive deployments could be away from home for as much as seven months a year. The isolation, long periods away and lengthy, often dangerous and stressful missions flown by ELINT crews bred a special closeness among them. This manifested itself in many ways, humor among them.

The RB-47H had a black radome nose as well as several other protuberances that made it stand out from the normal bomber versions. If we had to land at a destination other than our home base, we were usually diverted to another SAC base which had its own B-47 bomber Wings. It was not unusual to have a small crowd gather while we were parking, wondering what this slightly odd looking B-47 was. . . . What was the black nose used for? . . . Our response, many times, besides mentioning that the mission was classified, was that the black nose was used in refueling training for young pilots, in case we bumped into a tanker! Believe it or not, most who asked accepted that answer!

JACK KOVACS, PILOT, 55TH SRW, RB-47H/EB-47E(TT), 1959–1963

55th SRW EWO, Dave Johnson faces a panel full of dials, scopes, and knobs inside the cramped Crow compartment of an RB-47H. (*Dave Johnson*)

As far as is known, at least five Stratojets were fired on and damaged by Soviet fighters or antiaircraft defenses. All of these aircraft were RB-47s. Two of the three Stratojets which were actually shot down were RB-47Hs. As previously detailed, Bruce Olmstead and the crew of his RB-47H were shot down over the Barents Sea in July 1960. Another RB-47H was attacked and damaged near North Korea in 1965, and it is suspected that a second RB-47H was shot down over the Caspian Sea in the early 1960s.

The last RB-47H to be retired was also the last B-47 to serve in SAC and was flown to Davis Monthan AFB for storage on December 29, 1967. This aircraft (0-34296) did not sit in the Arizona desert for long, though. It was soon pulled out of storage to serve as a test bed for avionics systems being developed for the General Dynamics F/B-111. Flying into the

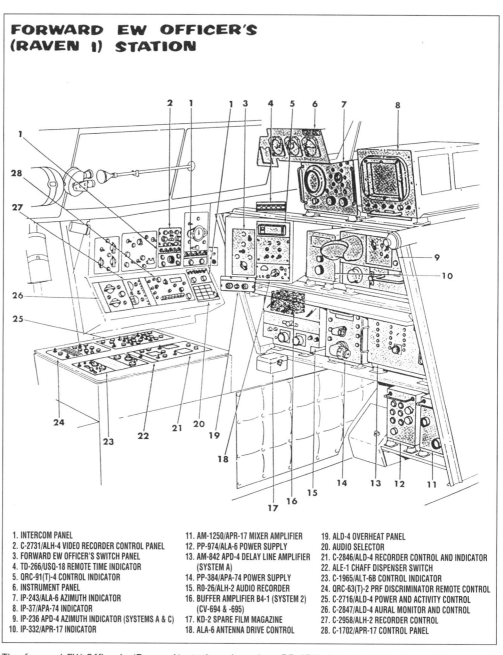

FORWARD EW OFFICER'S (RAVEN 1) STATION

1. INTERCOM PANEL
2. C-2731/ALH-4 VIDEO RECORDER CONTROL PANEL
3. FORWARD EW OFFICER'S SWITCH PANEL
4. TD-266/USQ-18 REMOTE TIME INDICATOR
5. QRC-91(T)-4 CONTROL INDICATOR
6. INSTRUMENT PANEL
7. IP-243/ALA-6 AZIMUTH INDICATOR
8. IP-37/APA-74 INDICATOR
9. IP-236 APD-4 AZIMUTH INDICATOR (SYSTEMS A & C)
10. IP-332/APR-17 INDICATOR
11. AM-1250/APR-17 MIXER AMPLIFIER
12. PP-974/ALA-6 POWER SUPPLY
13. AM-842 APD-4 DELAY LINE AMPLIFIER (SYSTEM A)
14. PP-384/APA-74 POWER SUPPLY
15. R0-26/ALH-2 AUDIO RECORDER
16. BUFFER AMPLIFIER B4-1 (SYSTEM 2) (CV-694 & -695)
17. KD-2 SPARE FILM MAGAZINE
18. ALA-6 ANTENNA DRIVE CONTROL
19. ALD-4 OVERHEAT PANEL
20. AUDIO SELECTOR
21. C-2846/ALD-4 RECORDER CONTROL AND INDICATOR
22. ALE-1 CHAFF DISPENSER SWITCH
23. C-1965/ALT-6B CONTROL INDICATOR
24. QRC-63(T)-2 PRF DISCRIMINATOR REMOTE CONTROL
25. C-2716/ALD-4 POWER AND ACTIVITY CONTROL
26. C-2847/ALD-4 AURAL MONITOR AND CONTROL
27. C-2958/ALH-2 RECORDER CONTROL
28. C-1702/APR-17 CONTROL PANEL

The forward EW Officer's (Raven 1) station aboard an RB-47H. A mass of equipment is present. (*Dave Johnsen*)

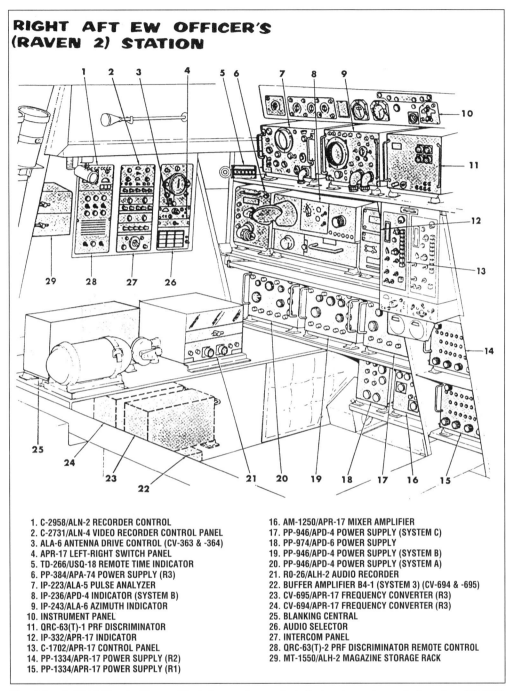

RIGHT AFT EW OFFICER'S (RAVEN 2) STATION

1. C-2958/ALN-2 RECORDER CONTROL
2. C-2731/ALN-4 VIDEO RECORDER CONTROL PANEL
3. ALA-6 ANTENNA DRIVE CONTROL (CV-363 & -364)
4. APR-17 LEFT-RIGHT SWITCH PANEL
5. TD-266/USQ-18 REMOTE TIME INDICATOR
6. PP-384/APA-74 POWER SUPPLY (R3)
7. IP-223/ALA-5 PULSE ANALYZER
8. IP-236/APD-4 INDICATOR (SYSTEM B)
9. IP-243/ALA-6 AZIMUTH INDICATOR
10. INSTRUMENT PANEL
11. QRC-63(T)-1 PRF DISCRIMINATOR
12. IP-332/APR-17 INDICATOR
13. C-1702/APR-17 CONTROL PANEL
14. PP-1334/APR-17 POWER SUPPLY (R2)
15. PP-1334/APR-17 POWER SUPPLY (R1)

16. AM-1250/APR-17 MIXER AMPLIFIER
17. PP-946/APD-4 POWER SUPPLY (SYSTEM C)
18. PP-974/APD-6 POWER SUPPLY
19. PP-946/APD-4 POWER SUPPLY (SYSTEM B)
20. PP-946/APD-4 POWER SUPPLY (SYSTEM A)
21. RO-26/ALH-2 AUDIO RECORDER
22. BUFFER AMPLIFIER B4-1 (SYSTEM 3) (CV-694 & -695)
23. CV-695/APR-17 FREQUENCY CONVERTER (R3)
24. CV-694/APR-17 FREQUENCY CONVERTER (R3)
25. BLANKING CENTRAL
26. AUDIO SELECTOR
27. INTERCOM PANEL
28. QRC-63(T)-2 PRF DISCRIMINATOR REMOTE CONTROL
29. MT-1550/ALH-2 MAGAZINE STORAGE RACK

The right aft EW Officer's (Raven 2) station and assorted electronic equipment aboard an RB-47H. (*Dave Johnson*)

early 1970s from Los Angeles Air Force Station, the aircraft was instantly recognizable by its F-111-like needle nose.

ERB-47H

The ERB-47H was an even more specialized ELINT platform than the RB-47H. Configured to carry only two Ravens in its Crow compartment, the space normally occupied by the Raven One in the RB-47H was used for additional equipment. The manner in which the ERB-47H was employed differed slightly as well.

> When the RB-47H picked up something of high interest that was new or different, the ERB would be tasked to go in and get more technical information on it. The ERB had a greater technical capability than the standard RB-47H. It was common practice to fly the two of them together and let the RB-47H go in and bait the systems to get them turned on. Then, the ERB would stand off and get the data. The ERBs flew many missions alone too but quite often we'd use them with the RB-47Hs as bait.
>
> **BRUCE BAILEY, EWO, 55TH SRW, 1958–1971**

Three ERB-47Hs (nos. 36245, 36246, 36249) were completed from among the last of the 35 RB-47Hs to be built. The ERB-47Hs were deployed all over the world, even operating from bases were no RB-47Hs were found. Crew duties for the Raven One and Two in the ERB-47H were somewhat similar to those for the three Ravens in the RB-47H. However, Crows qualified to fly in one or the other aircraft were not interchangeable. Switching between these aircraft would've required special training, and because of the small number of ERBs, only the most elite select crews were chosen to fly them. There were only about four crews qualified to fly the three ERBs.

The ERB-47H was gradually phased out along with the RB-47H as newer, more capable RC-135s took over the ELINT mission.

EB-47E(TT)

Another special variant to serve with the 55th SRW was the EB-47E(TT). This variant was quite different from the RB/ERB-47H ELINT Stratojets. Though it was also an intelligence

An RB-47H hooks up with a KC-135 in front of Mount McKinley, Alaska. There appears to be a shark's mouth just behind the black nose radome on the underside of the nose section. While artwork rarely appeared on B-47s of any variety, there were some aircraft that received adornment, at least temporarily. Artwork was for the most part forbidden, but when on TDY at far-flung bases around the world, 55th SRW crews did occasionally indulge themselves and a number of 55th aircraft wore some kind of nose art. (*William Henderson*)

gathering platform, the EB-47E(TT) went after a different type of information, telemetry intelligence (TELINT).

The 55th's EB-47E(TT)s were part of Detachment-4 at Incirlik Air Base, Turkey and flew the TELINT mission against two Russian launch sites. They collected telemetry data on the IRBM (Intermediate Range Ballistic Missile) facility at Kapustin Yar and the Soviet Space Center at Tyuratam. Unlike the other crews of the 55th, EB-47E(TT) crews were kept on alert, ready to scramble to be in place to intercept the telemetric data and communications that flowed during IRBM and space vehicle launches. Crews stood by two at a time, one as the primary alert crew and the other as the secondary or backup crew in case the primary aircraft experienced problems and had to abort within the first hour of the mission.

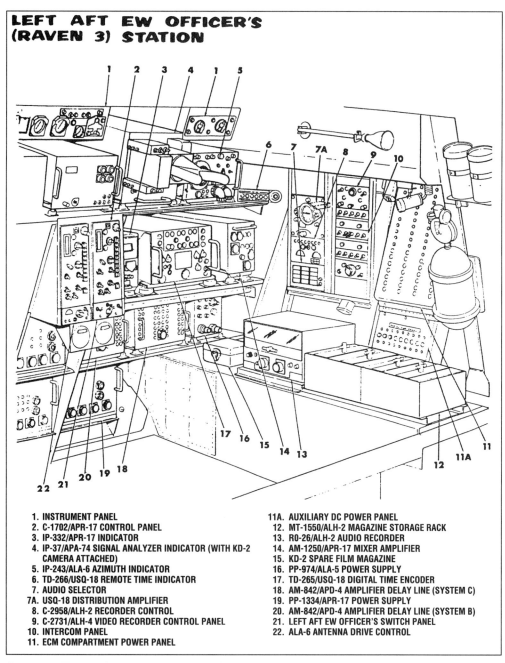

LEFT AFT EW OFFICER'S (RAVEN 3) STATION

1. INSTRUMENT PANEL
2. C-1702/APR-17 CONTROL PANEL
3. IP-332/APR-17 INDICATOR
4. IP-37/APA-74 SIGNAL ANALYZER INDICATOR (WITH KD-2 CAMERA ATTACHED)
5. IP-243/ALA-6 AZIMUTH INDICATOR
6. TD-266/USQ-18 REMOTE TIME INDICATOR
7. AUDIO SELECTOR
7A. USQ-18 DISTRIBUTION AMPLIFIER
8. C-2958/ALH-2 RECORDER CONTROL
9. C-2731/ALH-4 VIDEO RECORDER CONTROL PANEL
10. INTERCOM PANEL
11. ECM COMPARTMENT POWER PANEL
11A. AUXILIARY DC POWER PANEL
12. MT-1550/ALH-2 MAGAZINE STORAGE RACK
13. RO-26/ALH-2 AUDIO RECORDER
14. AM-1250/APR-17 MIXER AMPLIFIER
15. KD-2 SPARE FILM MAGAZINE
16. PP-974/ALA-5 POWER SUPPLY
17. TD-265/USQ-18 DIGITAL TIME ENCODER
18. AM-842/APD-4 AMPLIFIER DELAY LINE (SYSTEM C)
19. PP-1334/APR-17 POWER SUPPLY
20. AM-842/APD-4 AMPLIFIER DELAY LINE (SYSTEM B)
21. LEFT AFT EW OFFICER'S SWITCH PANEL
22. ALA-6 ANTENNA DRIVE CONTROL

One more diagram from an RB-47H Dash-One showing the left aft EW Officer's (Raven 3) station. (*Dave Johnson*)

The ERB-47H was distinguishable from the RB-47H by the chin antenna below its black nose radome. Note the bulge in the bomb bay section to accommodate the Crow compartment. (*National Archives*)

Leaving the runway at Incirlik, the EB-47E(TT) would climb toward one of the two facilities. At a predetermined location (above the Black Sea for Kapustin Yar or over Northeastern Iran for Tyuratam) the crew would begin an orbit at as high an altitude as possible, keeping station and collecting data until the launch sequences were complete. Launches were often delayed forcing the backup crew to be launched to go on station, replacing the primary crew at the limit of its endurance. Launches were often scrubbed as well, forcing another mission to be flown the next day. EB-47E(TT)s could also collect information on orbiting vehicles as they passed overhead and were occasionally sortied over the Pacific Ocean to collect data on the Soviet IRBM reentry points and target test landing areas.

Just three TTs were delivered. These aircraft were not purpose-built reconnaissance platforms like the RB/ERB-47Hs. Rather, they were B-47Es modified with the installation of a Crow capsule in the bomb bay like the Phase V ECM EB-47Es of the 376th BW. A totally different complement of receivers and recorders faced the two EWOs who sat side by side in the capsule and, aside from turning on their equipment at the proper time, there was little for the Crows to do.

The most noticeable modification to the EB-47E(TT) was the "armrest" or "towel rail" type antennas which ran along both sides of the fuselage below the cockpit. Later, smaller, more slender "chopstick" antennas appeared. However, the two antenna configurations were exactly the same. In the second configuration, the covers which housed the antennas were simply removed, it having been discovered that the antennas functioned properly without them. Either way, the antennas gave the aircraft a most unique look and once again provided opportunities for humor.

A head-on view of an ERB-47H. Its distinguishing chin antenna is clearly visible. Only three ERB-47Hs were built. (*National Archives*)

An ERB-47H flies in formation with a 55th SRW RB-47H. (*Dave Johnson*)

A very rare EB-47E(TT) on the ramp at Forbes AFB, Kansas in the early 1960s. This early (TT) sports the large antenna covers that were first used on the aircraft, giving it a very unique look. (*Jack Kovacs*)

Pilot Jim Nelson (in the middle) and the other four members of his crew stand in front of one of the three EB-47E(TT)s. These aircraft were used for the specialized TELINT mission. Note the large antenna covers which still house the antennas on this (TT). (*Jim Nelson*)

The EB-47E(TT) had large pontoonlike antennas on each side of the fuselage below the cockpit. On base in Turkey, we were known as members of the "secret SAC detachment." One evening, while sharing beers on the patios of the crew quarters, a group of F-100 fighter pilots started discussing the amount of weaponry they carried. Then, the discussion slowly shifted to their bragging that they "really knew what we did" on our secret missions. If we ever got in a bind with bad guys in fighters chasing us, all we had to do was give them a call and they'd help us out. After they'd mentioned this several times we reminded them of the big antennas on our fuselage and told them, "Thanks, but we don't really need help because if the bad guys fired a missile at us all we had to do was push the RTF (return to fighter) button and the missile would turn around and home in on the fighter that shot it (fighter pilot jaws drop). Most weren't sure whether or not to believe us . . . and we left it that way.

JACK KOVACS, PILOT, 55TH SRW, RB-47H/EB-47E(TT), 1959–1963

The EB-47E(TT) was phased out of service in 1967.

Conclusion

Though it seems a long time ago, the days of B-47s operating from SAC bases in the United States and overseas are not that far removed. Direct links to that period can be found within today's millennial Air Force. While newer more technologically advanced bombers like the B-1B and B-2 have come along, the B-52, only a generation apart from the B-47, still serves. Even less distant is the world which the Stratojet was so much a part of. It has been little more than a decade since the end of the superpower conflict, and though collective memory is often short, the image of Stratojets standing alert, ready to deliver a nuclear strike is one easily recalled. The legacy of the B-47 is still with us.

The B-47 is certainly one of the most significant aircraft of the twentieth century, both in terms of its influence on the course of large aircraft design and its impact on world history. It was the first truly successful jet bomber. Figures vary depending on the source, but it is generally agreed that no less than 2,042 Stratojets were built (counting the two XB-47 prototypes and the B-47C). The number of B-47s produced and the speed with which they were constructed is astonishing. Manufacture of the first production version of the aircraft, the B-47B, began in 1951. Less than six years later in February of 1957 the last B-47E rolled out the hangar doors at Boeing-Wichita. The B-47 program represents the largest bomber building effort since World War II. Why were so many built? Several factors played a role, including the onset of the Korean War and rising tensions between the East and West. The greatest impetus, however, came from the young Air Force's Strategic Air Command, or more precisely, its commander, General Curtis LeMay. It was his philosophy that quantity as well as quality enhanced deterrence, and though he was not terribly enthusiastic about the B-47 when it made its debut, he quickly realized that with the aid of aerial refueling, the Stratojet could make a powerful statement about American capabilities. In addition to being the most prolifically produced bomber since World War II, the Stratojet began the American policy of building strategic bombers exclusively for use by U.S. Armed Forces. B-47s never served operationally outside the United States (the CL-52 was a test bed solely) and none of the strategic bombers that followed it have ever been exported.

B-47s leaving contrails from their six J47 engines at high altitude. (*National Archives*)

In historical terms, the Stratojet made an undeniable contribution to the peace and security of our nation. During the height of the Cold War, the B-47 was among America's foremost strategic deterrents. Before the advent of ICBMs, it represented the country's primary delivery system for nuclear weapons. Nuclear capable B-47 units deployed to England, Western Europe, Africa, and Asia, all within minutes of targets inside Russia, doubtlessly had a measurable effect on Soviet policy makers. Furthermore, the advanced performance and capability of the new bomber forced the Soviets to design increasingly complex aircraft and weapons systems of their own in an expensive battle to keep pace with American technology.

The Stratojet was every bit as exotic as the B-2 is today when it first appeared in 1947. Advances in aerodynamics and electronics which were being discovered during World War II came to rapid fruition in the immediate postwar years and most of these were applied to the B-47. It was a groundbreaking aircraft which set trends in aircraft design which are still adhered to. But the Stratojet was more than innovative, it was also versatile. The Air Force pushed the B-47 to its limit, extracting every ounce of utility from the basic airframe. When the service had a new mission and cast around for an aircraft to fill the role, the B-47 frequently answered the call, performing duties it was never designed for. Through nearly 20 years of service the B-47 was a bomber, reconnaissance platform, flying test bed, hurricane hunter, trainer, and target. It performed all of these missions credibly and safely. Moreover, it was largely responsible for Boeing's tremendous success in the postwar era. Finally, as illustrated in this book, those who flew, crewed, and maintained the B-47 have, in general, nothing but praise for it. All acknowledge that it was a challenging aircraft to operate, but most remember their association with it fondly. This raises the question: Why has it so often been overlooked in the annals of aviation history? Why has it provoked so little comment?

Interestingly and perhaps tellingly, despite the thousands of Stratojets produced and the tens of thousands of men associated with it, the aircraft never acquired a nickname—affectionate or derogatory. In the often satirical world of aviation, many American and foreign navy and air force aircraft of equal or less significance than the B-47 acquired well-known handles. If someone mentioned a *Thud,* you knew they were talking about an F-105, a *Stoof* was a Grumman S2F, a *Fort* was a B-17, a *Scooter* an A-4 Skyhawk, a *Harley* an A-7 Corsair II, and the *Spad* or *Sandy* was an A-1 Skyraider. The Spitfire was a *Spit,* the Hawker Hurricane a *Hurry Box,* the C-130 a *Herk* or *Herky-Bird,* even the B-47's younger brother, the B-52, acquired the nickname *Buff.* Many more aircraft have been referred to in creative ways by aviators past and present, but surprisingly the B-47 is not among them.

On a more serious note, the Stratojet was perhaps a victim of its own success. The remarkable design spawned many more aircraft including two Boeing aircraft which are still in service, the KC-135 and B-52. While the KC-135, 707, and most of today's commercial aircraft are a legacy of the Stratojet, it is the B-52 which has been most responsible for eclipsing the B-47. No one who observed the first flight of the YB-52 in April 1952 could have anticipated that it would still be actively performing the mission it was designed for almost a half century later. No one would have guessed that an aircraft with the same 35 degree swept wings and a similar tandem gear arrangement to the B-47 would have dropped bombs in Vietnam, Desert Storm, and Allied Force. The Stratofortress is the direct descendant of the B-47, an aircraft built on the technology demonstrated and proven by the B-47. Developed hot on the heels of the Stratojet, the original YB/XB-52 borrowed even more from the B-47, including its tandem cockpit, pylon-mounted podded engines, and radar-directed tail guns.

Other factors have played a role in the relative obscurity of the B-47. While the B-52 has participated in every major conflict since Korea, the B-47 remained a virgin, never having seen combat in any armed conflict. Thus, it never gained a reputation as a battle-tested bomber. By the late 1950s the development of ICBMs had matured and the Stratojet was no longer the country's only or best system for delivering nuclear weapons. As a conventional bomber it was once again outclassed by its younger sibling which could carry more than twice its load of high explosives and carry them much farther. This is not to say that the B-47 didn't outperform the B-52 in some areas including top speed (especially at low altitude) and maneuverability. But by the early 1960s it was clear that technology's rapid march had passed the Stratojet by.

Finally, the environment and manner in which B-47s were operated may have contributed to their neglect by historians. Though the aircraft was glamorous when it first appeared, the mission it performed never really captured the public's attention. That was fine with SAC which had extremely tight security and generally shunned publicity about the specifics of its operations. Relatively few photos of the Stratojet exist when compared with the huge collections of images of other aircraft. Moreover, operational records of the aircraft are difficult to find and some, especially those which detail the operations of ELINT B-47s, have only recently been declassified. Consequently, access to information about the Stratojet has been limited. There is, however, a reservoir of knowledge that exists in the minds of the many men whose lives were touched by the B-47. While a small attempt has been made here to mine their collective memory much more information and history await collection.

The legacy of the B-47 is still with us in the B-52s and KC-135s which continue to fly in today's USAF, in the commercial airliners all around us which owe their basic designs to the Stratojet, in the contribution the B-47 made to winning the Cold War, and in the minds of the men who rose time after time from runways around the world to defend our nation's vital interests in the Boeing B-47 Stratojet.

B-47 Bomb Wings
Locations and Dates of Assignment

SAC B-47 Bomb Wings were divided among the 2nd, 8th, and 15th Air Forces. Throughout their history several of the units were transferred not only from location to location but between the three Air Forces. The following is a list of B-47 equipped Bomb Wings and the squadrons that comprised them, their locations, commands and dates of assignment.

Unit	Squadron	Location	Dates of Assignment
2nd BW (2nd AF)	20th BS 49th BS 96th BS 429th BS	Hunter AFB, GA	Nov. 25, 1953–Apr. 1, 1963 (transitioned to B-52s)
9th BW (15th AF)	2nd BS 19th BS 33rd BS 658th BS	Mtn. Home AFB, Idaho	Nov. 1, 1952–Mar. 15, 1963 (converted to SR-71)
19th BW (2nd AF)	28th BS 30th BW 93rd BW 658th BS	Pinecastle AFB, FL (transferred to Homestead AFB, FL, July 1956)	June 11, 1954–Jan. 1, 1961 (transitioned to B-52s)
22nd BW (15th AF)	2nd BS 19th BS 33rd BS 408th BS	March AFB, CA	Nov. 1, 1952–Mar. 15, 1963 (transitioned to B-52s)
40th BW (2nd AF) (15th AF) (2nd AF)	25th BS 44th BS 45th BS 660th BS	Smoky Hill AFB, KS (renamed Schilling AFB Jan. 1957) Forbes AFB, KS	May 28, 1952–Sept. 1954 (transferred to 2nd AF) Sept. 1954–June 20, 1960 June 20, 1960–Sept. 1, 1964 (deactivated)
43rd BW (15th AF)	63rd BS 64th BS 65th BS 408th BS	Davis–Monthan AFB, AZ	Sept. 25, 1953–Mar. 15, 1960 (converted to B-58)

(continued)

Unit	Squadron	Location	Dates of Assignment
44th BW (2nd AF)	66th BS 67th BS 68th BS 506th BS	Lake Charles AFB, LA (renamed Chenault AFB)	Apr. 25, 1953–June 15, 1960 (deactivated—became 44th SMW)
68th BW (2nd AF)	51st BS 52nd BS 656th BS 657th BS	Lake Charles AFB, LA (renamed Chenault AFB)	Oct. 25, 1953–Apr. 15, 1963 (transferred to 8th AF)
93rd BW (15th AF)	328th BS 329th BS 330th BS	Castle AFB, CA (SAC Training Wing)	Apr. 15, 1954–Feb. 1, 1955 (transitioned to B-52s)
96th BW (2nd AF)	337th BS 338th BS 339th BS 413rd BS	Altus AFB, OK	Nov. 18, 1953–Sept. 3, 1957 (transferred to 15th AF)
(15th AF)		Dyess AFB, TX	Sept. 4, 1957–Mar. 15, 1963 (transitioned to B-52s)
97th BW (15th AF)	340th BS 341st BS 342nd BS	Biggs AFB, TX	Apr. 1, 1955–Oct. 1, 1959 (transitioned to B-52s)
98th BW (15th AF)	343rd BS 344th BS 345th BS 415th BS	Lincoln AFB, NE	Oct. 15, 1954–Dec. 31, 1958 (transferred to 2nd AF)
(2nd AF)		Lincoln AFB, NE (redesignated 98th SAW, Feb. 1, 1964)	Dec. 31, 1958–June 25, 1966 (deactivated)
100th BW	349th BS 350th BS 351st BS 418th BS	Portsmouth AFB, NH (renamed Pease AFB in 1957)	Jan. 1, 1956–June 25, 1966 (transitioned to B-52s)
301st BW (2nd AF)	32nd BS 352nd BS 353rd BS 419th BS	Barksdale AFB, LA	June 20, 1953–April 15, 1958 (transferred to 8th AF)
(8th AF)		Lockbourne AFB, OH	Apr. 15, 1958–June 15, 1964 (transitioned to KC-135)
303rd BW (15th AF)	358th BS 359th BS 360th BS 427th BS	Davis–Monthan AFB, AZ	Jan. 20, 1953–June 15, 1964 (deactivated)
305th BW (15th AF)	364th BS 365th BS 366th BS 422nd BS	MacDill AFB, FL	April 15, 1952–Feb. 16, 1961 (transferred to 2nd AF)
(2nd AF)		Bunker Hill AFB, IN	Feb. 1961–Apr. 1, 1963 (converted to B-58)
306th BW (2nd AF)	367th BS 368th BS 369th BS 423rd BS	MacDill AFB, FL	April 2, 1951–Jan. 1, 1959 (deactivated—became 306th SW)
		MacDill AFB, FL	Jan. 1, 1959–Apr. 1, 1963 (transitioned to B-52s)

Unit	Squadron	Location	Dates of Assignment
307th BW (15th AF)	370th BS 371st BS 372nd BS 424th BS	Lincoln AFB, NE	Nov. 8, 1954–Dec. 31, 1958 (transferred to 2nd AF)
(2nd AF)		Lincoln AFB, NE	Dec. 31, 1959–June 25, 1965 (deactivated)
308th BW (2nd AF)	373rd BS 374th BS 375th BS 425th BS	Hunter AFB, GA	Sept. 25, 1952–June 25, 1961 (transferred to Plattsburgh AFB, NY, July 15, 1959) (deactivated)
310th BW (2nd AF)	379th BS 380th BS 381st BS 428th BS	Smoky Hill AFB, KS (renamed Schilling AFB)	May 10, 1954–Dec. 31, 1958 (transferred to 15th AF)
(15th AF)		Schilling AFB, KS	Dec. 31, 1958–Feb. 25, 1965
320th BW (15th AF)	441st BS 442nd BS 443rd BS 444th BS	March AFB, CA	July 23, 1953–Sept. 15, 1960
321st BW (2nd AF)	445th BS 446th BS 447th BS 448th BS	Pinecastle AFB, FL (renamed McCoy AFB)	Dec. 15, 1953–Jan. 1, 1959 (transferred to 8th AF)
(8th AF)		Pinecastle AFB, FL	Jan. 1, 1959–Oct. 25, 1961
340th BW (15th AF)	486th BS 487th BS 488th BS 489th BS	Whiteman AFB, MO	Aug. 8, 1953–Dec. 31, 1958 (transferred to 2nd AF)
(2nd AF)		Whiteman AFB, MO	Dec. 31, 1958–Sept. 1, 1963 (transitioned to B-52s)
341st BW (15th AF)	10th BS 12th BS 440th BS 491st BS	Dyess AFB, TX	Sept. 1, 1955–June 25, 1961 (deactivated—became 341st SMW)
376th BW (2nd AF)	512th BS 513th BS 514th BS 515th BS	Barksdale AFB, LA	Feb. 23, 1954–Nov. 14, 1957 (transferred to 8th AF)
(8th AF)		Lockbourne AFB, OH	Nov. 15, 1957–Mar. 15, 1965 (deactivated—became 376th SW)
379th BW (2nd AF)	524th BS 525th BS 526th BS 527th BS	Homestead AFB, FL	Nov. 1, 1955–Jan. 1, 1959 (transitioned to B-52s)
380th BW (2nd AF)	528th BS 529th BS 530th BS 531st BS	Plattsburgh AFB, NY (designated 380th SAW Sept. 15, 1964)	July 1, 1955–June 25, 1965 (transitioned to B-52s)
384th BW (2nd AF)	544th BS 545th BS 546th BS 547th BS	Little Rock AFB, AR	Aug. 1, 1955–April 2, 1966 (deactivated)

(continued)

Unit	Squadron	Location	Dates of Assignment
509th BW (15th AF)	343rd BS 661st BS 715th BS 830th BS	Walker AFB, NM	July 1, 1955–June 30, 1958 (transferred to 8th AF)
(8th AF)		Pease AFB, NH	July 1, 1958–April 2, 1966 (transitioned to B-52s)
4347th CCTW (2nd AF)	4347th TS 4348th TS 4349th TS 4350th TS	Wichita AFB, KS (renamed McConnell AFB)	July 1, 1958–June 15, 1963

Preserved Stratojets
Where to See Them

Though the majority of the over 2,000 Stratojets produced were scrapped, some remain. Many of these aircraft are on display in museums, at airports, air parks, air force bases, and other locations. This is an up-to-date list of the remaining B-47s and where to see them.

Type	Serial Number	Location
1. XB-47	46-00066	Octave Chanute Aerospace Museum (displayed as "2278")
2. B-47A	49-01901	Pima County Museum (nose only on display)
3. B-47B	50-00062	Mighty 8th Air Force Heritage Museum, Savannah, GA
4. B-47B	51-02075	Edwards AFB, CA
5. B-47B	51-02120	Whiteman AFB, MO
6. B-47B	51-02315	Grissom AFB, IN
7. WB-47E	51-02360	New England Air Museum, CT
8. B-47E	51-02387	Oklahoma City, OK
9. WB-47E	51-07066	Museum of Flight, Seattle, WA
10. B-47B	51-07071	Altus, OK
11. EB-47E-DT	52-00166	Castle Air Museum, CA (last flying B-47)
12. EB-47E	52-00410	Ellsworth AFB, SD (nose only on display)
13. EB-47E	52-00412	Linear Air Park, Dyess AFB, TX (displayed as "24120")
14. B-47E	52-00595	Little Rock AFB, AR
15. EB-47E	52-01412	SAC Museum, Offutt AFB, NE
16. B-47E	53-02104	Memorial Airport, Pueblo, CO
17. EB-47E	53-02135	Pima County Museum, AZ

(continued)

Type	Serial Number	Location
18. B-47E	53-02275	March AFB Museum, CA
19. B-47E	53-02276	Barksdale AFB, LA
20. EB-47E	53-02280	Wright-Patterson AFB Museum, OH
21. B-47E	53-02385	Former Plattsburgh AFB, NY
22. B-47E	53-04213	McConnell AFB, KS
23. RB-47E	53-04257	Tinker AFB, OK
24. RB-47H	53-04296	Eglin AFB Museum, FL
25. RB-47H	53-04299	Wright-Patterson AFB Museum, OH

Jan Tegler is a professional writer and broadcaster specializing in aviation. He is the author of numerous articles for domestic and international aviation publications on contemporary and historical subjects. He is also a private pilot and racecar driver.